RESEARCH IN MOLECULAR SPECTROSCOPY

ISSLEDOVANIYA PO MOLEKULYARNOI SPEKTROSKOPII

ИССЛЕДОВАНИЯ ПО МОЛЕКУЛЯРНОЙ СПЕКТРОСКОПИИ

Proceedings (Trudy) of the P.N. Lebedev Physics Institute

Volume 27

RESEARCH IN MOLECULAR SPECTROSCOPY

Edited by
Academician D. V. Skobel'tsyn

Authorized translation from the Russian

Springer Science+Business Media, LLC
1965

ISBN 978-1-4899-4940-0   ISBN 978-1-4899-4938-7 (eBook)
DOI 10.1007/978-1-4899-4938-7

The Russian text was published by Nauka in Moscow in 1964 for the USSR
Academy of Sciences as Volume XXVII of the Proceedings (Trudy) of the
P. N. Lebedev Physics Institute.

Исследования по молекулярной спектроскопии

Труды Физического института
Том XXVII

Library of Congress Catalog Card Number 65-14628

CONTENTS

INVESTIGATION OF INSTRUMENT DISTORTIONS AND METHODS OF CORRECTING FOR THEM IN INFRARED SPECTROSCOPY

G. G. Petrash

INTRODUCTION

The question of how to take into account the distortions introduced into a measured spectrum by the spectral apparatus, as well as the limitations imposed by those distortions, is one that arose a very long time ago. The first research on this problem was conducted by Lord Rayleigh, who published his well-known definition of resolving power [1] and developed a method of correcting for slit distortions that has not diminished in significance even to the present day [2]. Since then an enormous number of work has been published on the subject, to the extent that it would be impossible here to give a complete and comprehensive survey. We will attempt to cover the main results achieved in this area, devoting greater attention to those portions of the subject with which we have had occasion to come in direct contact.

It is a well-known fact that under certain conditions,* which are rarely violated and which we will henceforth consider to be satisfied (see, e. g., the conclusion of the survey article by Rautian [3]), the spectral distribution $f(\lambda)$ observed by a particular spectral instrument is described by an integral relation of the type

$$f(\lambda) = \int_{-\infty}^{\infty} \varphi(\lambda') \, a(\lambda - \lambda') \, d\lambda', \tag{0.1}$$

where $\varphi(\lambda')$ is the true distribution being measured, the function $a(\lambda)$ is called the instrument function of the given instrument and represents the observed spectral distribution that would be obtained with the given instrument in the investigation of monochromatic radiation. We will be working with distributions dependent on the wavelength λ. However, we will also use the same approach in considering distribution dependent on the frequency, wavelength, or coordinates in the recording reference frame. This is related to the fact that in deriving Eq. (0.1) we have already limited our discussion to relatively narrow spectral intervals, for which the transformation coefficients from one variable to another may be regarded as constant. In all that follows we will use only the variable λ.

It will be to our advantage [3] to normalize the instrument function over the area to unity:

$$\int_{-\infty}^{\infty} a(\lambda) \, d\lambda = 1. \tag{0.2}$$

This means that its effect has been reduced to a redistribution of energy in the spectrum; any factors essential to determining the absolute value of a spectral distribution, such as the transmissivity, luminosity, etc., will be treated separately if necessary.

Equation (0.1) is called the convolution (or faltung) of the functions $\varphi(\lambda)$ and $a(\lambda)$. Being a linear relation, the convolution has the properties of commutativity, associativity, and distributivity [3]. Its integral is equal to the product of the integrals of the functions $\varphi(\lambda)$ and $a(\lambda)$. In our normalization this implies that

---

*These conditions are dictated by the assumption that the entire spectral equipment complex comprises a linear system and that the instrument function remains unchanged over an interval of the same order of magnitude as its width.

$$\int_{-\infty}^{\infty} \varphi(\lambda)\, d\lambda = \int_{-\infty}^{\infty} f(\lambda)\, d\lambda, \tag{0.3}$$

i. e., the total energy of the observed distribution is equal to the total energy of the true distribution.

Taking the Fourier transform of both sides of Eq. (0.1), we obtain, in accordance with the theorem of the spectrum of the convolution [4],

$$F(\omega) = A(\omega)\Phi(\omega). \tag{0.4}$$

We will consistently use upper-case symbols to denote the Fourier transforms of the functions denoted by the lower-case counterparts:

$$F(\omega) = \int_{-\infty}^{\infty} f(\lambda)\, e^{-i\omega\lambda} d\lambda, \qquad \Phi(\omega) = \int_{-\infty}^{\infty} \varphi(\lambda)\, e^{-i\omega\lambda} d\lambda, \quad A(\omega) = \int_{-\infty}^{\infty} a(\lambda)\, e^{-i\omega\lambda} d\lambda. \tag{0.5}$$

If several independent instrument functions act in sequence on the measured distribution the total instrument function $a_{total}(\lambda)$ of the instrument complex as a whole is determined by the convolution of these instrument functions, and its Fourier transform is the product of the Fourier transforms of the individual instrument functions. In the ensuing sections we will differentiate between the instrument function of strictly spectral instrumentation (its optical part) $a(\lambda)$ and the instrument function of the registration system $k(\lambda)$. Then the Fourier transform of the total instrument function becomes

$$A_{tot}(\omega) = A(\omega)\, K(\omega),$$

and the Fourier transform of the observed distribution $h(\lambda)$ recorded at the output of an instrument in the registration system becomes

$$H(\omega) = A(\omega)\, K(\omega)\, \Phi(\omega). \tag{0.6}$$

In this demarcation we will not be specific with regard to the properties of the instrument function, designating it simply by $a(\lambda)$.

We turn once again to Eqs. (0.1) and (0.4). The search for the true distribution $\varphi(\lambda)$ is reduced to solution of the integral equation (0.1) or the algebraic equations (0.4), (0.6) and a subsequent Fourier transformation. In the latter case the solution can be written in the form

$$\Phi(\omega) = F(\omega)/A(\omega), \qquad \varphi(\lambda) = \frac{1}{2\pi} \int_{-\infty}^{\infty} \frac{F(\omega)}{A(\omega)} e^{i\omega\lambda} d\omega. \tag{0.7}$$

The implication is that in principle it is known. If we know $a(\lambda)$ and have measured $f(\lambda)$ the true distribution $\varphi(\lambda)$ can be obtained from Eq. (0.7).

However, to obtain $\varphi(\lambda)$ by this general method is a matter involving considerable computational difficulties, and for this reason it has not been used in practice so far. In its place a great many different methods have been proposed for the solution of Eq. (0.1), in each case relying on some form of simplification. In our discussions we will follow the established terminology in referring to the derivation of $\varphi(\lambda)$ from the observed funtion $f(\lambda)$ as reduction to an ideal instrument or, simply, reduction.

The proposed reduction methods fall generally into two classes.

1. The first includes methods of reduction presupposing a specific form for $\varphi(\lambda)$. In some case the form of $\varphi(\lambda)$ may be known from some additional sources of information (other measurements, theoretical considerations).

In some case the reduction problem is simplified. In fact, we are now able to compute the integral (0.1) by a more or less complicated analytical procedure. This enables us, granted in tabular form, to compare each $f(\lambda)$ with the specified $\varphi(\lambda)$. This means that the problem is solved in principle. In practice, however, such an approach succeeds only in the case of relatively simple functions $\varphi(\lambda)$ and $a(\lambda)$ depending on at most one or two parameters. For more complex functions computation of the integral soon becomes an ultimately problematical operation.

2. Approximate reduction methods constitute the second class. These methods are based on representation of the sought-after true distribution in the form

$$\varphi(\lambda) = f(\lambda) + \sum_i b_i L_i f(\lambda), \tag{0.8}$$

where L_i is some operator operating on $f(\lambda)$. S. G. Rautian [3] showed that such a representation is equivalent to expansion of the function $A^{-1}(\omega) - 1$ in a series of different functions $u_i(\omega)$:

$$A^{-1}(\omega) - 1 = \sum_i b_i u_i(\omega). \tag{0.9}$$

Expansion in a power series gives a representation of $f(\lambda)$ in terms of derivatives of $f(\lambda)$, which has been obtained by different authors independently [5-9]. Expansion in functions $u_i = \sin^i q\omega/2$ leads to a representation of $\varphi(\lambda)$ in terms of finite differences [2, 10-14]. Other representations are also possible [15-18]. We will not discuss them further, referring the reader instead to the review article by Rautian [3].

We see from the foregoing brief survey that there are not inherent shortcomings in reduction methods. Two problems remain open to discussion: 1) the existence and uniqueness of a reduction; 2) the comparison of various reduction methods from the point of view of their benefits in accuracy.

The first problem has been investigated in [19-24], wherein it is shown that when a number of reasonable conditions are satisfied there is a unique solution to Eq. (0.1). This conclusion derives from the assumption that the functions $f(\lambda)$ and $a(\lambda)$ are known with absolute precision and implies that if the measurement accuracy is increased without limit there always exists in principle the possibility of reproducing the true distribution $\varphi(\lambda)$. It follows from this that the problem of the accuracy of spectral measurements, like the problem of the limiting characteristics of spectral instruments (for example, the resolving power), cannot in general be treated correctly without allowing for the errors of measurement.* This means, then, that the various reduction methods and spectral instruments cannot be compared without taking into account the random errors of measurement.

Consequently, the problem of accurate characteristics of spectral instruments and the influence of measurement errors becomes one of great conceptual importance.

The problem of the role of random errors is treated the most systematically by L. A. Khalfin [22, 25, 26]. His approach may be succintly described as follows. A series of conceptual true distributions $\varphi_i(\lambda)$(hypotheses) is considered. For each of these the corresponding theoretical observed distribution $f_i(\lambda)$ is calculated. Inasmuch as the measured function $f(\lambda)$ contains random errors (noise) none of the $f_i(\lambda)$ will necessarily agree with $f(\lambda)$ in general. Proceeding from the differences between $f_i(\lambda)$ and $f(\lambda)$ and knowing the error probability distribution, one can evaluate the probability that the true distribution is $\varphi_i(\lambda)$. This gives us a measure of the most probable $\varphi_i(\lambda)$ and the mean error in determining $\varphi(\lambda)$. In the event that the a priori probabilities φ_i are known beforehand, measurement of $f(\lambda)$ enables us to reevaluate these probabilities. In order for our measurement of $f(\lambda)$ to refine the information about $\varphi(\lambda)$ we must know at least the second moment of the error distribution.

---

*To avoid confusion from the outset, we will henceforth distinguish between two kinds of measurement error, systematic errors or systematic distortions, a term denoting deviations of the observed contour from the true one due to any sort of instrument function effects, and random errors, denoting errors in the measurement of the observed contour due in our case to various random processes originating in the measuring apparatus.

This approach makes it possible to obtain a number of important general postulates (for example, concerning the conceptual necessity of allowing for noise, the necessity of knowing the statistical characteristics of the noise, the role of a priori information, etc.). However, its direct application for the practical computation of $\varphi(\lambda)$ is restricted, since it means calculating the integral (0.1) for each φ_i, an operation that is exceedingly difficult if φ_i and $a(\lambda)$ are at all complex and can scarcely be accomplished today in a reasonable amount of time, even with the help of computer technology. It is appropriate, therefore, to adopt another approach. The first step is to perform an approximate reduction and to obtain a more or less corrected distribution $\varphi_a(\lambda)$ to the extent that the residual systematic distortions need no longer be regarded. The resultant distribution $\varphi_a(\lambda)$ will contain the transformed random errors of $f(\lambda)$, whose statistical characteristics can be calculated, knowing the initial values. $\varphi_a(\lambda)$ can now be treated by the method of Khalfin,* wherein it is no longer required each time to compute f_i, it being sufficient to compare $\varphi_a(\lambda)$ and $\varphi_i(\lambda)$. This makes it possible to carry out the difficult solution of Eq. (0.1) just once. Following this approach, we arrive at the problem of transforming the errors in reduction. This problem has been treated in its general aspect by S. G. Rautian [20], whose work is briefly summarized as follows.

The presence of measurement errors is reflected in the fact that, instead of (0.1), we must now consider the more general expression

$$f(\lambda) = \int_{-\infty}^{\infty} \varphi(\lambda') a(\lambda - \lambda') d\lambda' + \xi(\lambda), \tag{0.10}$$

where $\xi(\lambda)$ describes the random errors of the measurement. The Fourier transform of $f(\lambda)$ becomes

$$F(\omega) = \Phi(\omega) A(\omega) + X(\omega). \tag{0.11}$$

As we have already seen, the reduction process amounts to dividing $F(\omega)$ by $A(\omega)$ or by its approximate expression; hence it affects the errors and in some way transforms them. If we characterize the errors by their correlation function $\psi(\Delta\lambda) = \overline{\xi_\lambda \xi_{\lambda+\Delta\lambda}}$, the spectral power of the error will be determined by the Fourier transform of $\psi(\Delta\lambda)$:

$$\Psi(\omega) = \int_{-\infty}^{\infty} \psi(\Delta\lambda) e^{-i\omega\Delta\lambda} d \cdot (\Delta\lambda). \tag{0.12}$$

If we perform an exact reduction according to Eq. (0.7) the error in the final result will be [27]

$$\overline{\Delta\varphi^2} = \frac{1}{2\pi} \int_{-\infty}^{\infty} \Psi(\omega) |A(\omega)|^{-2} d\omega. \tag{0.13}$$

It is readily apparent that this integral diverges, since the function $A(\omega)$ tends to zero as ω increases, while $\psi(\omega)$ usually does not change much over a broad interval of ω [3]. The situation as just described is illustrated in Fig. 1, which shows the functions considered so far.

Thus exact reduction with errors present in $f(\lambda)$ generates an infinite error in the final result and is therefore impracticable in general. Only an approximate reduction is possible, for which the function $A^{-1}(\omega)$ is replaced by a function $G(\omega)$ approximating it over some finite interval of ω. The reduction result contains both systematic errors due to incomplete reduction and transformed errors of the observed distribution. Consequently, the approximate reduction is equivalent to an instrument function that reduces the systematic error at the expense of increasing the random error of $f(\lambda)$. The more accurate the reduction the greater will be

---

*It is difficult at the present time to calculate even $\varphi_a(\lambda)$ according to Khalfin's method, because curves with noise are processed in spectroscopy largely by "averaging by eye." The application of more modern processing techniques clearly must await adequate developments in computer technology.

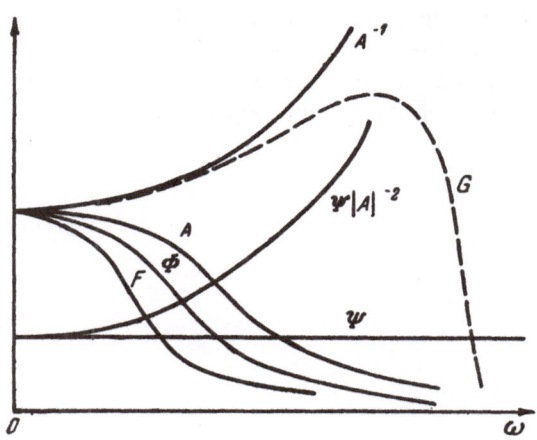

Fig. 1. Form of the pertinent functions.

the role of the random errors. Given this situation, it is natural that the question of an optimum reduction should arise [3], i.e., of a reduction for which maximum measurement precision is obtained in measuring a particular spectral characteristic. This problem is intimately connected with the problem of optimum conditions of measurement. The essence of the matter is that in the majority of situations the experimenter can vary both the width of the instrument function and the magnitude of the random errors, where a reduction in width of the instrument function and the concomitant reduction in systematic error tends usually to increase the random error of $f(\lambda)$. A familar example is the case when narrowing the slits of a monochromator diminishes the systematic slit distortions but at the same time decreases the intensity of the incident light, thus increasing the relative errors of the measurement. It can be shown that in the extreme cases of very narrow and very wide slits the total error is large, while the best accuracy is obtained at some intermediate slit width.

This example, which is very typical of spectroscopic measurements, shows us how important it is to be able to make a proper choice of experimental conditions.

Despite this fact, so far only the very first steps have been taken in the quest for optimum conditions of measurement.

We have outlined above Rautian's analysis of the general problems involved in the reduction and choice of conditions for measurement. Beyond this we are only able to mention calculations of optimum measurement conditions for a number of specific problems [20] carried out by the same author, and a few papers [28-30] touching on the choice of an optimum absorption region. (We will return to all of these papers in more detail in the appropriate sections.) Consequently, further work is needed in the search for optimum conditions for the solution of specific spectroscopic problems.

The analysis presented above with regard to instrument distortions was concerned with optical spectroscopy. The applicability of the methods developed here and the results obtained with them are by no means confined exclusively to this area, as a similar situation occurs in many other areas of physical measurement, for example in recording high-speed and short-lived processes in electrical linear systems, in nuclear spectroscopy, x-ray photography, in radioastronomy, in the theory of optical devices, and others. This is related to the fact that any real physical instrument tends to "diffuse" the fine structure of a measured distribution, and that the accuracy of measurements is always finite. Since the quantitative relations describing instrument distortions are compatible in many cases, certain general results will prove applicable to other areas of physical measurement, despite differences in the specific form of the instrument functions and measurement errors. The concrete results suitable for application in practical measurements, however, can usually be obtained only over a comparatively narrow interval. In view of the essential differences in the properties of spectral instruments, radiation detectors, and recording systems in particular, it is impossible to set down any practical conclusions universally applicable to all instruments.

In the present study we intend not only to consider some general aspects of correcting for instrument distortions but also to apply the results of the general analysis toward specific problems of practical interest. Due to the overwhelming diversity of practical problems we find it necessary to stay within a given range of problems. As already noted, the greatest importance attaches to the compensation of instrument distortions in infrared spectroscopy. Of all the many and variegated problems of infrared spectroscopy, particular attention is bestowed upon the problem of obtaining absolute spectral characteristics in cases of considerable scientific and practical significance.

We have attempted to give all calculations requiring specific assumptions as to the attributes of the spectral instruments for conditions approximating as nearly as possible the conditions typical of modern infrared spectroscopy. Nevertheless, the methods developed below, as well as the many results obtained in the corresponding consideration of the specific features of spectral equipment, can be used equally well for measurements in other regions of the spectrum.

CHAPTER I

OPTIMUM CONDITIONS FOR MEASUREMENT IN THE ABSENCE OF REDUCTION [31-37]

It will first of all be necessary to sharply delineate the class of problems under discussion. Inasmuch as we are presently concerned with the choice of optimum conditions for measurement it is essential to indicate just what measurements we have in mind. In actual point of fact quantitative measurements of the most diverse sort may be involved in the investigation of spectra. The goal may be, for example, to measure the frequency of lines or bands or their relative intensities, or the problem may be one of resolution, etc.

It is important to note that to each of the various kinds of measurement corresponds in general a particular set of optimum conditions, such that in analyzing the problem of choosing proper measurement conditions one must state what parameters are being measured and what sort of measurement procedure is involved.

We could have considered the measurement of concentration of a particular component in a mixture in terms of its infrared spectrum or the utilization of a calibration graph relating the sought-after concentration of the component with, say, the peak intensity of a definite band. In this case the systematic instrument distortions essentially produce no errors since they are accounted for by the calibration graph. However, they decrease the value of the observed absorption and, hence, influence the accuracy and sensitivity of the method. The errors in this case are determined by the random errors introduced in recording the spectrum (at this point we are ignoring the effects of other components in the mixture). In this circumstance one might pose the question as to what conditions will yield the highest experimental accuracy in measuring the concentration of a given component, i. e., what are the optimum conditions for measurement in this specific analysis situation.

We will be concerned with another problem, namely the measurement of the absolute values of the spectrum parameters, i.e., the values of parameters which are exempt from the influence of the apparatus.

In this connection, any departures from the true spectrum brought in during the measurement process will be regarded as errors. This problem is one of the most acute in the present state of infrared spectroscopy.

We will be concerned below with the following parameters of absorption bands: optical density, optical density at peak absorption, bandwidth, and to a certain extent band configuration.

Measurements of the integral optical density are not considered. Errors in the measurement of this quantity are connected chiefly with the uncontrollable influence of overlapping adjacent bands. Correction for this type of error is beyond the scope of the problems with which we are concerned.

In considering the accuracy and optimum conditions of measurement in this and ensuing chapters we will take into account the following instrument distortions.

1. Systematic errors related to the influence of the instrument function of the spectral equipment and manifested in the well-known lowering of the spectral peaks and broadening of the bands.

2. Random errors related to uncompensated random processes in the instrument. Noise in the radiation detection section provides the main contribution here.

3. "Mechanical errors," i.e., distortions related to imperfection of the registration system, their magnitude being characterized usually by the concept "a class of instrument precision" (as they are regarded in a number of instances).

These three types of distortion play a fundamental role, since they cannot be eliminated. We will devote our attention below to such relatively tractable distortions as fluctuations of the solid angle of the instrument and sensitivity of the radiation detector as a function of wavelength, the effect of scattering in the instrument, etc.

The problem of measurement errors and the choice of conditions for measurement in the absence of reduction is of both theoretical and practical importance. We will treat it from the practical point of view.

In many studies (dealing with, for example, measurement of the spectra of unique or unusual substances or spectra obtained under special conditions, at low or high temperatures, etc.) the results of direct measurements that have not been subjected to any sort of reduction are given. It is important therefore to state the methods used to estimate the accuracy attained in such measurements and elucidate the conditions under which more accurate data can possibly be obtained. Such estimates are necessary also for further analysis of the application of reduction methods, since in order to judge the advantage of this or that method of reduction it is necessary to find out how much greater the accuracy of the given method is than the accuracy of the measurements without reduction.

From the theoretical point of view the investigation of the accuracy of measurements without reduction essentially corresponds to the case of small systematic errors. The latter admits of a comparatively straightforward and graphic analysis of systematic error and of the overall problem of optimality in analytical form. This fact is extremely valuable, because one of the prime difficulties in analyzing systematic errors is directly related to the fact that the effect of the instrument function is not readily visualized. Proceeding from the general equation (0.1) it is very difficult to form a quantitative notion of how a given instrument function will operate on bands of various widths and configurations or how the effects of different instrument functions will differ from one another. Furthermore, in practical calculations it is usually necessary to assume a definite form for $a(\lambda)$ and $\varphi(\lambda)$. Any other form for $a(\lambda)$ and $\varphi(\lambda)$ renders it difficult to predict the result. In the analysis of small systematic errors, however, it is possible to obtain a simple analytical expression for a very general case; depending on the form of $a(\lambda)$ and $\varphi(\lambda)$, in this expression only the values of definite parameters vary, so that the effect of a change in form can be estimated immediately. In addition, simplification of the expressions makes it possible to include in the analysis such effects as the influence of the thickness of the investigated layer of substance on the configuration and width of the bands, permits simultaneous inclusion of the effects of the instrument function and response of the registration system, etc. Allowance for these details, which in general is an ultimately complex problem, enables one to arrive at a number of conclusions as to the role of any particular factors and, specifically, to demonstrate that some of them cannot be taken into account in the general case.

It is clear from the above that the analysis of systematic distortions and the methods for obtaining maximum accuracy in spectroscopic measurements must begin right at the analysis of measurements without subsequent reduction, i.e., with the assumption of small systematic errors.

§1. Systematic Errors

As we have already perceived (see the Introduction), the effect of all the instrument distortions can be described by the equation

$$H(\omega) = \Phi(\omega) A_{\text{tot}}(\omega). \tag{I.1}$$

Our assumption of small systematic errors implies that $H(\omega)$ differs only slightly from $\Phi(\omega)$, i.e., that in the interval of ω where $\Phi(\omega)$ is significantly different from zero $A_{\text{tot}}(\omega)$ differs only slightly from one. We will try replacing $A_{\text{tot}}(\omega)$ in this interval of ω by an approximate expression in order to simplify the result. The total instrument function of the equipment is the result of the sequential operation of the instrument function of the optical section of the equipment, i.e., the monochromator, and the instrument function of the registration system. It is convenient to separate these two independent instrument functions in order to differentiate between the separate dependences of the errors on the parameters of the monochromator and registration system. As will be evident from an analysis of the reduction methods (see Chapt. II), this demarcation is of fundamental importance, in that the noise amplitude is related in different ways with the parameters of these two instrument functions.

Let us denote by $A(\omega)$ the Fourier transform of the instrument function of the monochromator $a(\lambda)$, by $K(\omega)$ the Fourier transform of the instrument function of the registration system $k(\lambda)$. Ordinarily the spectrum to be investigated is scanned in time, and the instrument function of the registration equipment describes inertial distortions that arise in scanning. In this case the function $k(\lambda) = k(vt)$, where t is the time, v the scan rate, and $K(\omega)$ is simply the frequency characteristic of the registration system expressed as a function of ω.

In the general case the instrument functions $a(\lambda)$ and $k(\lambda)$ are asymmetrical and, accordingly, $A(\omega)$ and $K(\omega)$ are complex:

$$A(\omega) = |A(\omega)| e^{-i\eta\omega}, \qquad K(\omega) = |K(\omega)| e^{-i\theta\omega}. \tag{I.2}$$

Our approach will be to expand these expressions in a power series with respect to ω about the point $\omega = 0$. We recall that $|A(\omega)|$ and $|K(\omega)|$ are even functions, and the phases $\eta(\omega)$ and $\theta(\omega)$ are odd. Further, we know that the instrument functions must be normalized over the area to unity, i.e.,

$$A(0) = K(0) = 1.$$

In accordance with this we will assume that the following expansions are possible:

$$\left. \begin{aligned} |A(\omega)| &= 1 - a_2\alpha^2\omega^2 + a_4\alpha^4\omega^4 + \ldots, \\ |K(\omega)| &= 1 - b_2\tau^2v^2\omega^2 + b_4\tau^4v^4\omega^4 + \ldots, \end{aligned} \right\} \tag{I.3}$$

$$\left. \begin{aligned} \eta(\omega) &= \eta_1\omega + \eta_3\omega^3 + \ldots, \\ \theta(\omega) &= \theta_1\omega + \theta_3\omega^3 + \ldots \end{aligned} \right\} \tag{I.4}$$

Stopping with the first nonvanishing approximation, we obtain

$$A(\omega) \simeq (1 - a\alpha^2\omega^2) e^{-i\omega\eta}, \qquad K(\omega) \simeq (1 - b\tau^2v^2\omega^2) e^{-i\omega\theta}. \tag{I.5}$$

In these expressions α and τv are parameters determining the width of the distributions $|A(\omega)|$ and $|K(\omega)|$, respectively. These parameters have the dimensions of λ and specify the scale of $a(\lambda)$ and $k(\lambda)$ on the λ axis. It is convenient, for example, to adopt the characteristics normally used: α, the width of the instrument function $a(\lambda)$; τ, the time constant, or effective time constant of the registration system. For simplicity the subscripts associated with a, b, η, and θ have been omitted in Eqs. (I.5). The minus sign in front of the first term of the expansion was chosen to comply with the fact that the instrument function broadens the contour, i.e., $A(\omega)$ diminishes with increasing ω. Hence the quantities a and b are positive.

Limiting ourselves to the same approximation for $H(\omega)$, we now obtain

$$H(\omega) = \Phi(\omega) e^{-i\omega(\eta+\theta)} \{1 - a\alpha^2\omega^2 - b\tau^2v^2\omega^2\}. \tag{I.6}$$

We could also expand the exponential factor in front of the braces in a power series in ω, but, as will be explained later, it is more convenient to keep the expression in this form. It is known that the factor $e^{-i\omega(\eta+\theta)}$ represents a shift by the amount $\eta + \theta$, while multiplication by $(i\omega)^n$ means taking the n-th derivative of the corresponding function of λ (see, e.g., [4]). In this regard, taking the inverse Fourier transform, we find

$$h(\lambda) \simeq \varphi(\lambda - \eta - \theta) + \varphi''(\lambda - \eta - \theta) \{a\alpha^2 + b\tau^2v^2\}. \tag{I.7}$$

If we had included in the expansions (I.3) and (I.4) terms of higher order this expression would have contained higher derivatives. Expansion of the phase factor $e^{-i\omega(\eta+\theta)}$ in a power series in ω would lead to the appearance of the first derivative and variation of the coefficients in front of the other derivatives. This would detract greatly from the visual interpretability of the result, since it is important to isolate the shift of the entire contour as a whole. The shift of the contour as a whole can, as a matter of fact, be accounted for quite easily. Futhermore, in practice it is almost always eliminated in measurements because the position of the lines and bands in a spectrum is usually measured after an appropriate calibration of the instrument with respect to wavelength by some known spectrum. If the calibration is performed with the same registration system the shift is automatically eliminated because, as evident from (I.7), it is the same for all bands.

In the description of systematic errors in [38-40] expressions are obtained proportional to the first power of α or the first derivative of $\varphi(\lambda)$. It is readily seen that these errors correspond, correct to terms of order ω^2 and higher, to linear terms, i.e., the shift.

Departing from this, we will assume that the shift is eliminated by calibration, thereby disregarding the associated errors. Consequently, in the first nonvanishing approximation the systematic errors are proportional to the second derivative at a given point of the contour and the square of the width of the instrument functions:

$$h(\lambda) = \varphi(\lambda) + \varphi''(\lambda)\{a\alpha^2 + b\tau^2 v^2\}. \tag{I.8}$$

We note that the method given here for approximately describing the systematic error is allied with approximate reduction methods. As shown by S. G. Rautian [3], the various methods of approximate reduction essentially utilize an expansion of the function $A^{-1}(\omega) - 1$ in various functional series. Thus, a power series expansion leads to a correction expression in terms of the derivatives of the observed contour [5-9, 41, 42]; an expansion in trigonometric functions leads to an expression in finite differences (Rayleigh method, Bracewell method [2, 14]), etc. In the first approximation all of the functions used are normally well described by a square-law expression. This means that the errors accounted for in the first approximation of these methods correspond to the systematic errors that we have obtained herein.

For our present purposes it will be useful to separate further the dependence of the width of the measured contour in explicit form. We note in doing this that, given a fixed form for $\varphi(\lambda)$, the n-th derivative can be written as follows:

$$\varphi^n(\lambda) = U_n(\lambda)\varphi_0/\gamma^n, \tag{I.9}$$

where φ_0 is the peak values of $\varphi(\lambda)$, defining the ordinate scale, γ is the width of the contour, defining its scale on the λ axis, and U_n is a factor representing the n-th derivative of a contour with the same shape but with a peak height and width of unity, i.e., depending only on the shape of $\varphi(\lambda)$. Appropriating this, we find

$$h(\lambda) = \varphi(\lambda) + \varphi_0 r(\lambda)\{a\alpha^2/\gamma^2 + b\tau^2 v^2/\gamma^2\}. \tag{I.10}$$

Appearing in this expression are the natural dimensionless ratios α/γ and $\tau v/\gamma$, which determine the systematic error. Only the factors a, b, and r vary as a function of the form of the functions $a(\lambda)$, $k(\lambda)$, and $\varphi(\lambda)$. These factors are determined by the relation

$$\varphi''(\lambda) = \varphi_0 r(\lambda)/\gamma^2. \tag{I.10\bar{a}}$$

We now consider the limits of applicability of the resultant equation. No restrictive assumptions were made with regard to $\varphi(\lambda)$ except, of course, the existence of the corresponding derivatives of $\varphi(\lambda)$, so that the equation is essentially applicable for any measured contour. As for the instrument functions, before applying Eq. (I.10) one must be sure that the expansions (I.3) and (I.4) do in fact exist. For infrared instruments this is usually so. To be more specific, we will look into the most frequently encountered instrument functions.

1. The instrument function of a single slit is

$$a(\lambda) = \begin{cases} 1/S & \text{for } |\lambda| \leqslant S/2, \\ 0 & \text{for } |\lambda| > S/2, \end{cases} \qquad A(\omega) = \frac{\sin \frac{\omega S}{2}}{\frac{\omega S}{2}}, \qquad (I.11)$$

where S is the spectral width of the slit.

$$A(\omega) \simeq 1 - \frac{\omega^2 S^2}{24} + \frac{\omega^4 S^4}{1920} - \cdots, \qquad a = \frac{1}{24} \simeq 0.04167, \qquad \eta = 0.$$

2. For two slits of width S_1 and S_2

$$A(\omega) = \frac{\sin \frac{\omega S_1}{2}}{\omega S_1/2} \cdot \frac{\sin \frac{\omega S_2}{2}}{\omega S_2/2} \simeq 1 - \frac{1}{24}(S_1^2 + S_2^2). \qquad (I.12)$$

3. For slits of the same width

$$A(\omega) \simeq 1 - \frac{1}{12}\omega^2 S^2, \qquad a = 1/12 = 0.0833, \qquad \eta = 0. \qquad (I.12a)$$

4. An instrument function of Gaussian form is

$$a(\lambda) = \frac{2\ln 2}{\alpha \sqrt{\pi}} e^{-\frac{4\ln 2 \lambda^2}{\alpha^2}}.$$

Here and below α is the half-width (between half-peak points).

$$A(\omega) = e^{-\frac{\omega^2 \alpha^2}{16\ln 2}} \simeq 1 - \frac{1}{16\ln 2}\omega^2 \alpha^2 + \cdots, \qquad a = \frac{1}{16\ln 2} \simeq \frac{1}{11.09} = 0.09017, \quad \eta = 0. \qquad (I.13)$$

5. An exponential instrument function is*

$$a(\lambda) = \frac{\ln 2}{\alpha} e^{-\frac{2\ln 2}{\alpha}|\lambda|}.$$

$$A(\omega) = \frac{1}{1 + \omega^2 \alpha^2/(2\ln 2)^2} \simeq 1 - \frac{1}{(2\ln 2)^2}\omega^2 \alpha^2,$$

$$a = \frac{1}{(2\ln 2)^2} = 0.5205, \qquad \eta = 0. \qquad (I.14)$$

We see that the instrument functions listed above fit into our scheme; the next two, however, do not:†

6. A dispersion type instrument function is

$$a(\lambda) = \frac{2/\pi\alpha}{1 + 4\lambda^2/\alpha^2}. \qquad A(\omega) = e^{-\alpha|\omega|} \simeq 1 - \alpha|\omega|. \qquad (I.15)$$

---

*According to [43-45], this form is characteristic of the instrument function of a photosensitive layer.
†This has to do with the fact that these functions possess slowly varying wings.

7. The diffraction instrument function is

$$a(\lambda) = \frac{1}{S_0}\left[\frac{\sin \pi\lambda/S_0}{\pi\lambda/S_0}\right]^2. \qquad A(\omega) = 1 - \frac{1}{2\pi}S_0|\omega|; \qquad (I.16)$$

$S_0 = \lambda f/D$ is the normal slit width.

We now wish to examine the characteristics of registration systems. It is known [46] that for systems with lumped parameters the frequency characteristic can be written in the form

$$K(\omega) = \frac{P_1(i\omega)}{P_2(i\omega)}, \qquad (I.17)$$

where P_1 and P_2 are polynomials of $i\omega$. If we assume that $K(0) = 1$ and the registration system has an integrating capability, as is typical of most instruments, then for such a system the expansions (I.3) and (I.4) are always admissible and, consequently, it fits into our scheme. It is usually assumed that the registration system is equivalent to a simple integrating RC circuit with time constant τ. In this case

$$K(\omega) = \frac{1}{1 + i\omega\tau\upsilon},$$

$$|K(\omega)| = \frac{1}{\sqrt{1 + \omega^2\tau^2\upsilon^2}} \simeq 1 - \frac{1}{2}\omega^2\tau^2\upsilon^2, \qquad (I.18)$$

$$\eta(\omega) = \arctan \omega\tau\upsilon \simeq \omega\tau\upsilon, \quad \eta = \tau\upsilon.$$

It is readily apparent that a combination of n noninteracting RC circuits yields

$$K(\omega) = \prod_{j=1}^{n} K_j(\omega) \simeq \left[1 - \frac{\omega^2}{2}\sum_{j=1}^{n}\tau_j^2\upsilon^2\right]e^{-i\omega\sum_{j=1}^{n}\tau_j\upsilon},$$

$$b = \frac{1}{2}, \quad \tau_{equ}^2 = \sum_{j=1}^{n}\tau_j^2, \quad \eta = \upsilon\sum_{j=1}^{n}\tau_j.$$

L. A. Gribov investigated the response characteristics of a registration system of two-beam infrared instruments [40]. The differential equation obtained for the system gives

$$K(\omega) \simeq \left[1 - \omega^2\left(\frac{1}{K_s^2} - \frac{2}{K_s}(\tau_a - \tau_{mo})\right)\right]e^{-i\frac{\omega}{K_s}}, \qquad (I.19)$$

where K_s is the system gain, τ_a and τ_{mo} are the time constants of the amplifier and processing motor, respectively.

Our entire analysis is made on the assumption that the systematic error is small. We will examine what this means in actual practice. Let us proceed by measuring the dispersion band with an instrument whose instrument function has the form of an isosceles triangle of width S [$a(\lambda)$ for a pure slit; two slits of equal width]. For this case

$$A(\omega) \simeq 1 - \frac{1}{12}\omega^2 S^2 + \frac{1}{360}\omega^4 S^4 - \ldots$$

We require that for ω such that $\Phi(\omega)$ is $\frac{1}{10}$ of Φ_0 the third term with $\omega^4 S^4$ will be one-tenth the second term. This yields $S^2/\gamma^2 \simeq \frac{1}{7}$, corresponding to $\Delta\varphi/\varphi_0 \approx \frac{1}{10}$, i.e., 10% error. Consequently, our equations are sufficiently accurate as long as the systematic errors do not exceed 10%. If we take into consideration the fact that under optimum conditions the random error is approximately equal to the systematic error and add the error due to lag in the registration system, this limit will correspond to about 20% of the total error. In the optical density this quantity will correspond to even greater error, on the order of 30 to 50%.

This rough estimate shows that our equations are valid practically for all cases when it is sensible to work without reduction because when the error is too large, as a rule, the measurements are of no practical value anyway. In this case it is necessary to use an instrument with better characteristics or to apply reduction methods.

§2. Optimum Conditions for the Measurement of Optical Density

We now make use of the expressions obtained for the systematic error to find the optimum conditions for the operation of spectral instrumentation. The spectroscopic problems solved on a given instrument can differ in their nature, so that, accordingly, the optimum operating conditions of the instrument will in general also differ. In this chapter we will consider one of the most important and current problems of infrared spectroscopy, measuring the true spectral distribution of the absorption coefficient of a substance, i.e., obtaining the spectral characteristics of the substance that are independent of the instrumentation.

We will use the Bouguer-Lambert-Beer law for relating the absorptivity x or transmissivity $T = 1 - x$ to the optical density D:

$$T\,(\lambda) = I_\lambda(\lambda)/I_0 = 1 - x\,(\lambda) = e^{-D(\lambda)} = e^{-\varepsilon(\lambda)cd}). \tag{I.20}$$

Here I_0 is the unperturbed light intensity, $I(\lambda)$ is the intensity of the light as it passes through the sample, $\varepsilon(\lambda)$ is the absorption coefficient of the sample investigated, c is the concentration of the investigated substance, d is its thickness. We will not consider the measurements errors of c and d inasmuch as they are not related to the spectral instrumentation. Hence we need only consider measurements of the optical density D.

Let us establish the connection between the errors in D and the absorptivity x. From the equation $D = -\ln(1 - x)$ we obtain

$$\frac{\overline{\Delta D^2}}{D^2} = \frac{\overline{\Delta x^2}}{(1-x)^2 \ln^2(1-x)}. \tag{I.21}$$

We now consider the error in the measurement of x. We need three readings in order to determine it (Fig. 2); a reading l_0 to the line of total transmission, a reading l to the line marking the measured absorption, and a reading l_n to the line of total absorption. We then obtain

$$I_0 = l_n - l_0, \quad \varphi\,(\lambda) = l - l_0$$

and, finally,

$$x = \varphi\,(\lambda)/I_0 = (l - l_0)/(l_n - l_0). \tag{I.22}$$

For the error we find

$$\Delta x = \Delta l/I_0 - x\,\Delta l_n/I_0 - (1 - x)\,\Delta l_0/I_0. \tag{I.23}$$

Since the measurement errors of the indicated three lines are independent,

$$\overline{\Delta x^2} = \overline{\Delta l^2}/I_0^2 + x^2\overline{\Delta l_n^2}/I_0^2 + (1 - x)^2\overline{\Delta l_0^2}/I_0^2. \tag{I.24}$$

The lines of total absorption and total transmission are found at a special position since they do not have a narrow structure. As will be evident later on, the accuracy of the measured spectral distribution is determined by the width of the spectrum structure and the precision characteristics of the spectral instrument. Consequently, the same instrument can be used to measure the lines of total absorption and total transmission, which do not have structure, far more accurately than the spectrum of the sample. This can be accomplished, for example, by broadening of the slit width and an increase in the time constant of registration system in

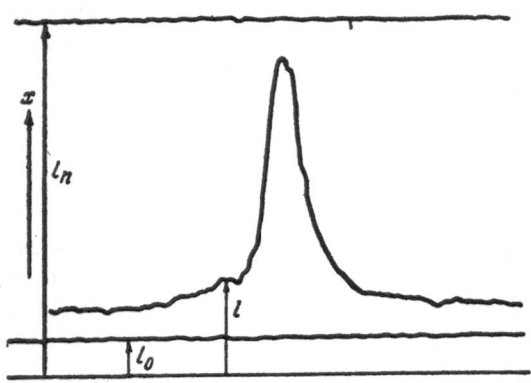

Fig. 2. Typical reading in the measurement of absorption.

recording these lines. But even if this is not done, the measurement accuracy of these lines will still be higher, since they can be recorded over a relatively large interval and then averaged, even if by no other technique than "by eye." These arguments show that it is always feasible in practice to do something so that the error in measuring the lines of total transmission and total absorption will be small in comparison with the error in recording a line of the sample spectrum. Proceeding from this, we will not bother to account for the measurement error involved in the lines of total absorption and transmission, dropping the corresponding terms in Eq. (I.24).

Let us now look at the error $\overline{\Delta l^2}$ in measuring a line of the sample spectrum. It is composed of systematic and random errors. For a description of the systematic error we will use the results of the preceding section, drawing on a dissertation by Petrash [31] for our account of the random error (see also the Conclusion). We will not consider here the "mechanical error," since in the absence of reduction it will be significant only when the measurement accuracy is near the limit, and in this case the error analysis is irrelevant. We assume accordingly that

$$\frac{\overline{\Delta l^2}}{I_0^2} = \frac{E}{\tau S_1^2 S_2^2} \, .$$

(see the Conclusion). The parameter E characterizes the instrument noise and is equal to the mean square noise amplitude relative to I_0^2 for $\tau = 1$, $S_1 = S_2 = 1$. Taking both kinds of error into account, we obtain

$$\overline{\Delta x^2} = x_0^2 r^2(\lambda) \left\{ a\,\frac{\alpha^2}{\gamma_x^2} + b\,\frac{\tau^2 v^2}{\gamma_x^2} \right\} + \frac{E}{\tau S_1^2 S_2^2} \qquad (I.25)$$

and for the optical density error

$$P^2 = \frac{\overline{\Delta D^2}}{D^2} = \left[\frac{1}{(1-x)\ln(1-x)} \right]^2 \left(x_0^2 r^2 \left\{ a\,\frac{\alpha^2}{\gamma_x^2} + b\,\frac{\tau^2 v^2}{\gamma_x^2} \right\}^2 + \frac{E}{\tau S_1^2 S_2^2} \right). \qquad (I.26)$$

In these equations S_1 and S_2 are the widths of the incoming and outgoing slits of the monochromator, E is a characteristic of the instrument determining the noise amplitude, x is the absorptivity at that point of the band contour where the optical density D is measured, and x_0 is the peak absorption. The factor

$$M(x) = \left[\frac{1}{(1-x)\ln(1-x)} \right]^2,$$

as we have seen, appears in (I.26) as the result of going over from the absorption error Δx^2 of (I.25) to the relative optical density error, since the variables D and x are related by the nonlinear relation (I.20). However, P^2 depends not only on the absorptivity x but on the absorptivity x_0 as well, since the second derivative at that point of the contour where D is measured, and hence the systematic error, depend on x_0. In this regard, there appears in front of the braces the factor $x_0^2 r^2(\lambda)$, where $r(\lambda)$, as we have seen, represents the λ dependence of the second derivative of a given shape of contour but with unit x_0 and γ. This is not confined, however, to the dependence of P^2 on x_0. As a matter of fact, since the instrument functions operate on the contour in absorption, the width γ_x of the absorption band contour, which depends on the thickness of the measured sample layer and shape of the contour, figures into this expression. Transformation to the width γ_D of the contour in optical density, which is a characteristic of the substance and does not depend on the thickness of the layer, can be done according to the relation

$$\gamma_D = \gamma_x N(x_0).$$

The form of the function $N(x_0)$ depends on the form of $\varphi(\lambda)$ (see §3). The factor $r(\lambda)$ also depends on the form of $\varphi(\lambda)$, and this form changes as x_0 varies. It can be shown, however, that the variations of the form of $\varphi(\lambda)$ are small up to $x_0 = 0.8$. Since, as will be explained, it is unsuitable to work with large x_0, we will not take into account the variation of $r(\lambda)$ as a function of x_0.

With all this uppermost in mind, we find

$$P^2 = \frac{\overline{\Delta D^2}}{D^2} = M(x) \left(x_0^2 N^4(x_0) r^2 \left\{ a \frac{\alpha^2}{\gamma_D^2} + b \frac{\tau^2 v^2}{\gamma_D^2} \right\}^2 + \frac{E}{\tau S_1^2 S_2^2} \right). \tag{I.27}$$

As we shall see, the relative error in measuring the optical density depends on certain characteristics of the instrument: E, a, b, and on the following parameters which are at the disposal of the experimenter: S_1, S_2, τ, v, d. By varying the parameters we should find those conditions for which P will be a minimum. We note at once that there is no minimum with respect to v; only the systematic error depends on v and monotonically at that; the smaller the scan rate the less will be the error. This means that the scan rate must be chosen on the basis of other arguments. (This problem is discussed below.) With respect to the remaining parameters, however, a minimum does exist. So that the resultant equation can be used to find the minimum of P we must be still more specific about the relation between α and S.

As we have already mentioned, in infrared spectroscopy the instrument function is determined mainly be slits. We will examine in detail the case when the instrument function is determined exclusively by slits, winding up with the result for the case of small nonslit-type corrections to the width of the instrument function.

Equation (I.27) involves two slits S_1 and S_2, so that the choice of optimum conditions also includes the choice of ratio between the widths of the incoming and exit slits. In order to circumvent the cumbersome equations we will bypass the analysis of the problem and merely cite its result, which is very simple: $S_1 = S_2$ (this agrees with the strandard construction when the slits are equal, as for example in the case of IKS-11 and IKS-14 infrared spectrometers). This result has been obtained on the bais of other arguments by Toporets[47].

In accordance with the above, we let $\alpha = S = S_1 = S_2$ and obtain

$$P^2 = \frac{\overline{\Delta D^2}}{D^2} = \left[\frac{1}{(1-x) \ln(1-x)} \right]^2 \left(x_0^2 N^4(x_0) r^2 \left\{ a \frac{S^2}{\gamma_D^2} + b \frac{\tau^2 v^2}{\gamma_D^2} \right\}^2 + \frac{E}{\tau S^4} \right). \tag{I.28}$$

For convenience we introduce the natural dimensionless parameters

$$z = \frac{S}{\gamma_D}; \quad y = \frac{\tau v}{\gamma_D}, \quad U = \frac{Ev}{\gamma_D^5}, \quad M(x) = \left[\frac{1}{(1-x) \ln(1-x)} \right]^2. \tag{I.29}$$

Then

$$P^2 = M(x)(x_0^2 N^4 r^2 \{az^2 + by^2\}^2 + U/yz^4). \tag{I.30}$$

To aid in finding the minimum we have

$$\frac{\partial P^2}{\partial z} = M(x) \left(2x_0^2 r^2 N^4 2az \{az^2 + by^2\} - \frac{4U}{yz^5} \right) = 0,$$

$$\frac{\partial P^2}{\partial y} = M(x) \left(2x_0^2 r^2 N^4 2by \{az^2 + by^2\} - \frac{U}{y^2 z^4} \right) = 0. \tag{I.31}$$

From these two equations we establish a connection between the optimum values of z_m and y_m:

$$y_m = \frac{z_m}{2}\sqrt{\frac{a}{b}}. \tag{I.32}$$

This relation depends only on the connection between α and S and does not depend in the least on either x, x_0, or any assumptions relative to N, r, or $\overline{\Delta l_n^2}$, $\overline{\Delta l_0^2}$. It determines the relation that must exist at optimum conditions between $\tau_m \nu$ and S_m, provided the instrument function of the monochromator is determined only by the slits. This relation contains only parameters that depend on the properties of the instrument, namely on the form of the instrument function and the transfer chracteristic of the registration system. From this it is possible to obtain the relations that must exist between the various kinds of error at optimum:

$$\frac{P_m(\text{syst. } S)}{P_m(\text{syst. } \tau)} = 4, \tag{I.33}$$

i.e., at optimum conditions the systematic error due to the instrument function is four times the error due to the response of the instrument system. Moreover, the following relation is satisfied:

$$\frac{P_m^2(\text{syst.})}{P_m^2(\text{rand.})} = \frac{5}{4}. \tag{I.34}$$

The systematic error at optimum is approximately equal to the random error.

These last two relations do not depend at all on either the properties of the instrument or on the investigated spectrum. They are simply a consequence of the fact that S and τ enter into the expressions for the systematic and random errors in the corresponding degrees.

From Eqs. (I.31) we also find

$$z_m = \left(\frac{8U}{5a^2 r^2 N^4 x_0^2 \sqrt{a/b}}\right)^{1/9}, \tag{I.35}$$

$$P_m^2 = 2^{-\frac{8}{3}}\, 5^{5/9}\, 9 r^{10/9}\, N^{20/9}\, a^{5/9}\, b^{2/9}\, U^{\frac{4}{9}} M(x)\, x_0^{10/9}. \tag{I.36}$$

As is evident from these equations, the optimum slit width $z_m = S_m/\gamma_D$ and the minimum value of the total relative error P_m^2 depend both on the properties of the instrument and on the properties of the measured contour. We recall that the parameters a and b depend only on the form of the instrument function and the form of the frequency characteristic of the registration system, hence are constant for a given instrument. The choice of a suitable absorption value will be considered later; it will become apparent that x_0 can be predetermined. The form of the function $N(x_0)$, as well as the value of r, will depend to a certain extent on the shape of the measured contour.* The most significant dependence will be borne by the parameter $U = E\nu/\gamma_D^5$, which then essentially determines the capability of the instrument with respect to measuring a band of width γ_D. What is the sense of this parameter? If we call forth the formula for the random error and recognize that $\gamma_D/\nu = t_\gamma$ is the time to record the bandwidth it is then readily apparent that the parameter

$$U = \frac{E\nu}{\gamma_D^5} = \frac{E}{t_\gamma \gamma_D^4}$$

represents the mean square random error of measurement on the given instrument at a slit width $S = \gamma_D$ and time constant $\tau = t_\gamma$. This fundamental parameter, as we will have frequent occasion to perceive as the

---

*This will all be illustrated in specfic examples at the end of the chapter.

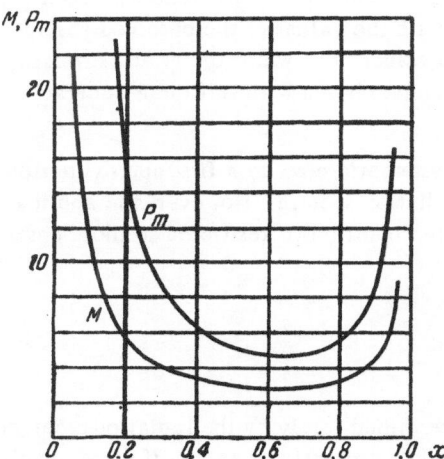

Fig. 3. Dependence of the relative error in measuring the optical density on the absorption.

discussion unfolds, plays a decisive role in all problems associated with the compensation of instrument distortions. This betrays the fact that the limiting capabilities of a spectral instrument are always governed in the final analysis by its precision characteristics.

We now turn to the problem of choosing the scan rate v. As we have seen, the only possibility for reducing P_m for the given instrument necessarily involves reducing v, i.e., increasing the measurement time. This situation is typical of those instances in which the measurement accuracy is confined to random processes. This is the case in the problem at which we are now looking. It is known that in such circumstances the problem of attainable measurement accuracy can only be properly stated with a specified time of measurement. This forces us into a consideration of the problem of optimum conditions and of the attainability under these conditions of the measurement accuracy with a fixed total measurement time, i.e., for a given spectrum at fixed v. It is possible, of course, to state the converse problem: At what peak scan rate can a given measurement accuracy of a definite spectrum be obtained on a given instrument? Equation (I.36) answers this problem for our experimental conditions.

Had we decided to let $\alpha = S + c$, where c is a small nonslit correction [20], we would have obtained the following optimum conditions:

$$y_m = \tau_m v = \frac{1}{2} \sqrt{\frac{a}{b} z_m \left(z_m - \frac{c}{\tau_D} \right)},$$

$$\frac{P_m (\text{syst. } \alpha)}{P_m (\text{syst. } \tau)} = 4 \left(1 + \frac{c}{S} \right), \qquad \frac{P_m^2 (\text{syst.})}{P_m^2 (\text{rand.})} = \frac{5}{4} + \frac{c}{S + c}. \qquad (I.37)$$

The equations for z_m and P_m^2 in this case turn out to be quite bulky, and we will not devote space to them. We will be content to note that small nonslit additives to the width of the instrument function do not significantly alter the conditions of measurement.

Let us now consider the dependence of the total error on the thickness of the layer d. Throughout our equations appear the absorptivity values x and x_0, which depend on d. This is more convenient, since it is ordinarily the absorption that is recorded on the instrument. We therefore reduce the choice of layer thickness to the problem of deciding on the absorptivity; we will consider the absorption dependence of the error. As we have seen, the optimum conditions for z and y are determined for any x and x_0 by Eqs. (I.32) and (I.35), but the actual values of z_m, y_m, and P_m depend on x and x_0, although not very strongly. Consequently, the choice of measurement conditions must begin with a selection of the layer thickness such that the measured values of x and x_0 are obtained within the required limits.

The dependence of P_m^2 on x, x_0 is expressed as follows:

$$P_m^2 \sim M(x) \, x_0^{10/\circ} N^{20/\circ} (x_0) = x_0^{10/\circ} N^{20/\circ} (x_0) \left[\frac{1}{(1 - x) \ln (1 - x)_1} \right]^2. \qquad (I.38)$$

The major contributing factor here is M(x). Its graph is illustrated in Fig. 3, with the optical density measured at the peak $x = x_0$. The relation $P_m(x_0)$ in this case is shown in the same figure, and $N(x_0)$ was calculated according to Eq. (I.44) assuming a dispersion type contour for $x(\lambda)$. As we shall see, the regions of minimum error for these curves are quite consistent. This means that we can also choose the value of x for the case $x \neq x_0$ within the same limits, provided only that x does not differ too appreciably from x_0.

It is necessary to point out that all of our equations involve x, the true absorption, whereas in practice it would be expeditious to choose the experimental conditions starting with the values of the observed absorption. Since, however, there exists a rather broad interval of admissible values of x, while the actual errors in the absorption are assumed to be small, the necessary layer thickness can be chosen with sufficient accuracy, using the observed values of the absorption.

We note in conclusion that Eqs. (I.7)-(I.10), which describe the systematic error to a first approximation, may be applied with equal success, both to absorption spectra and to radiation spectra. However, the applicability to radiation spectra of the results obtained with regard to optimum conditions is restricted to those cases when the following relation is satisfied:

$$\frac{\overline{\Delta l^2}}{l_0^2} = \frac{E}{\tau S^4} .$$

It was assumed in the derivation of the latter that the noise level is determined solely by the radiation receiver, and that the intensity of the light transmitted through the monochromator is proportional to S^2. If these conditions are met the results obtained above can also be applied to radiation spectra, as long as we take into account that the absorption dependence vanishes.

§3. Optimum Conditions for Measuring the Bandwidths

In §2 we explained the optimum conditions for measuring one of the most widely encountered problems of infrared spectroscopy, measurement of the optical density. It is very often necessary at the same time, however, to determine the width of the spectral bands. It is instructive, therefore, to calculate the error in measuring this characteristic of the absorption bands, to understand the optimum conditions for its measurement, and to compare these with the optimum conditions for measurement of the optical density.

Normally we are interested in the width of the band contours in the optical density $D(\lambda)$ because this quantity is directly connected with the properties of the substance investigated. It is customary in practice, however, to record the contours of the absorption bands $x(\lambda)$, and it is more convenient to measure the width for this contour.

For changing from the width of the absorption contour γ_x to the width of the optical density contour we use a method suggested by V. N. Smirnov [48]. After having demonstrated that the shape of the $x(\lambda)$ contour differs only slightly from the shape of $D(\lambda)$ as long as D is not too large (D < 1.7) and then taking advantage of this hypothesis, Smirnov arrived at the equation

$$\gamma_D = \gamma_x N (x_0). \tag{I.39}$$

Here the function $N(x_0)$ depends on the absorption at the band peak x_0, its form being determined by the shape of the band.

It can be shown, however, that this equation is equally valid for any x_0, and that the postulated agreement of the forms of $x(\lambda)$ and $D(\lambda)$ is not a necessary condition for its fulfillment inasmuch as it simply characterizes the ratio of scales on the λ axis. In fact, we can let $D[\lambda/(\gamma_D/2)]$ describe the shape of the optical density contour, where γ_D is the width of this contour at the height $\varkappa D_0$. We will consider the width at the height $\varkappa D_0$ in general, bearing in mind the possibility that the width may not be determined at the half-power points, as in the case of a curve with background.

By definition

$$D \left(\frac{\lambda}{\gamma_D /2} \right) = - \ln \left[1 - x \left(\frac{\lambda}{\gamma_x /2} \right) \right]. \tag{I.40}$$

Here $x[\lambda/(\gamma_x/2)]$ characterizes the shape of the absorption contour, γ_x the width of this contour at the height $\varkappa x_0$. Hence

$$D\left(\frac{\gamma_x}{\gamma_D}\right) = -\ln\left(1 - \varkappa x_0\right).$$
(I.41)

The resultant equation represents an implicit relation between γ_x and γ_D. It is clear that this relation is always described by Eq. (I.39). If in the identity (I.30) we substitute $\lambda = \gamma_D/2$ we then obtain

$$\varkappa D_0 = -\ln\left[1 - x\left(\frac{\gamma_D}{\gamma_x}\right)\right].$$
(I.42)

This expression also contains an implicit relation between γ_x and γ_D, except that now the form of the function $x[\lambda/(\gamma_x/2)]$ varies with x_0. If we assume that this form is invariant we then arrive at Eq. (I.39), except that now the function $N(x_0)$ has another form.

Let us examine some special cases.

1. Dispersion band:

$$D\left(\frac{\lambda}{\gamma_D/2}\right) = \frac{D_0}{1 + \dfrac{4\lambda^2}{\gamma_D^2}\dfrac{1-\varkappa}{\varkappa}}, \qquad N(x_0) = \left[\frac{\ln\left(1 - \varkappa x_0\right)}{\ln\left(\dfrac{1 - x_0}{1 - \varkappa x_0}\right)}\right]^{1/2}\sqrt{\frac{1-\varkappa}{\varkappa}}.$$
(I.43)

But if we assume that

$$x\left(\frac{\lambda}{\gamma_x/2}\right) = \frac{x_0}{1 + \dfrac{4\lambda^2}{\gamma_x^2}}, \qquad \varkappa = \frac{1}{2},$$
(I.44)

then [48]

$$N(x_0) = (1 - x_0)^{1/4}.$$

2. Gaussian band:

$$D\left(\frac{\lambda}{\gamma_D/2}\right) = D_0 e^{\frac{-4\ln\varkappa\lambda^2}{\gamma_D^2}}, \quad N(x_0) = \left[\frac{\ln\varkappa}{\ln\dfrac{\ln\left(1 - \varkappa x_0\right)}{\ln\left(1 - x_0\right)}}\right]^{1/2}.$$
(I.45)

If we let

$$x\left(\frac{\lambda}{\gamma_x/2}\right) = x_0 e^{\frac{-4\ln\varkappa\lambda^2}{\gamma_x^2}},$$

then

$$N(x_0) = \left[\frac{1}{\ln\varkappa}\ln\frac{1 - (1 - x_0)^\varkappa}{x_0}\right]^{1/2}.$$
(I.46)

Figure 4 shows these functions for the case $\varkappa = \frac{1}{2}$. The solid curves represent the functions $N(x_0)$ obtained with the assumption of a definite form imparted to $D(\lambda)$, the dashed curves assume a definite form for $x(\lambda)$. Curve 1 corresponds to a band with dispersion configuration, curve 2 to a Gaussian configuration. For small x_0 it is reasonable to expect the dashed and solid curves to fall together.

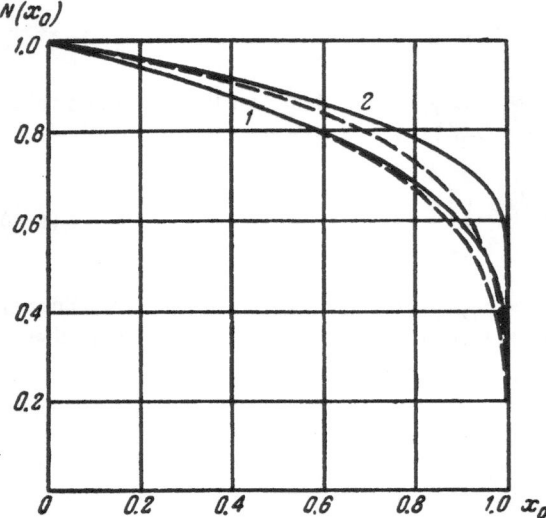

$N(x_0)$

Fig. 4. Relation between widths of absorption and optical density bands.

Let us now consider the error in measuring the width γ_D:

$$\frac{\Delta\gamma_D}{\gamma_D} = \frac{\Delta\gamma_x}{\gamma_x} + \frac{\Delta N}{N}. \qquad (I.47)$$

We will consider the nonreduction case. This means that we simply measure the width of the observed absorption contour, i.e., we need in essence to solve the equation

$$h\,(\delta_\pm) = \varkappa h_0. \qquad (I.48)$$

Here $h(\lambda)$ is the observed amplitude of the absorption line, $\delta_\pm$ are the points of intersection of the line $\varkappa h_0$ with the contour $h(\lambda)$. The observed width δ in this is case is defined as

$$\delta = \delta_+ - \delta_-.$$

Making use of Eq. (I.8) for $h(\lambda)$, we write Eq. (I.48) in the expanded form

$$\varkappa h_0 = \varkappa\varphi_0 + \varkappa\varphi_0''\{aS^2 + b\tau^2 v^2\} + \varkappa\Delta h_0 = \varphi\,(\delta_\pm) + \varphi''\,(\delta_\pm)\times\{aS^2 + b\tau^2 v^2\} + \Delta h_\pm = h\,(\delta_\pm). \quad (I.49)$$

The terms Δh_0 and $\Delta h_\pm$ describe the random errors in measuring the peak and at the height $\varkappa h_0$. Below we will analyze this equation for δ_+.

We will consider the error in width to be small, letting $\delta_+ = \gamma_+ + \Delta_+$ and expanding (I.49) in a series, stopping of course with just the first approximation. Then

$$\varkappa\varphi_0 + \varkappa\varphi_0''\{aS^2 + b\tau^2 v^2\} + \varkappa\Delta h_0$$

$$= \varphi\,(\gamma_+) + \varphi'\,(\gamma_+)\,\Delta_+ + \varphi''\,(\gamma_+)\{aS^2 + b\tau^2 v^2\} + \Delta_+\varphi''\{aS^2 + b\tau^2 v^2\} + \Delta h_+.$$

By stipulation $\varkappa\varphi_0 = \varphi(\gamma_\pm)$. With this in mind we obtain

$$\Delta_+ = \frac{[\varkappa\varphi_0'' - \varphi''\,(\gamma_+)]\{aS^2 + b\tau^2 v^2\} + \varkappa\Delta h_0 - \Delta h_+}{\varphi'\,(\gamma_+) + \varphi'''\,(\gamma_+)\{aS\} + b\tau^2 v^2\}}. \qquad (I.50)$$

The numerator comprises an expression describing the error in measuring h_0 and $h(\delta_\pm)$, the denominator describes the derivative of the observed curve and translates the ordinate error to the error in width. The second term in the denominator describes the difference between this derivative and the derivative of the true contour due to the influence of the instrument function. It is known, however, that at the half-power points the second derivative is very small, hence this terms must also be small. The analysis of typical examples has shown that it can be neglected.

Let us now investigate the second term in (I.47):

$$\frac{\Delta N}{N} = \frac{1}{N}\frac{dN}{dx_0}\Delta x_0. \qquad (I.51)$$

22

Substituting (I.50) and (I.51) into (I.47) and recognizing that for the error Δ_- an expression analogous to (I.50) will be obtained, we get

$$\frac{\Delta \gamma_D}{\gamma_D} = \frac{2\left[r_0 \varkappa - r\left(\frac{\gamma_x}{2}\right)\right]\{az^2 + by^2\}\, N^2 + \frac{2\varkappa \Delta h_0}{\varphi_0} - \frac{\Delta h_+}{\varphi_0} - \frac{\Delta h_-}{\varphi_0}}{p\,(\gamma_x/2)} + \frac{x_0}{N}\frac{dN}{dx_0}\left[r_0\{az^2 + by^2\}\, N^2 + \frac{\Delta h_0}{\varphi_0}\right]. \quad (I.52)$$

We have introduced at this point the notation

$$p\,(\lambda) = \frac{\gamma\varphi'(\lambda)}{\varphi_0}, \qquad r\,(\lambda) = \frac{\gamma^2\varphi''(\lambda)}{\varphi_0}. \quad (I.53)$$

Assuming, as is the custom, that the random measurement errors are independent at different points of the contour, we obtain

$$\mathscr{E}^2 = \frac{\overline{\Delta\gamma_D^2}}{\gamma_D^2} = L\,(x_0)\left[V^2\,(x_0)\,\{az^2 + by^2\}^2 + \frac{U}{yz^4}\right]. \quad (I.54)$$

Here the functions $L(x_0)$ and $V^2(x_0)$ depend only on x_0 and have the form

$$L\,(x_0) = \frac{\left[2\varkappa + \varkappa p\,(\gamma/2)\frac{x_0}{N}\frac{dN}{dx_0}\right]^2 + 2}{x_0^2 \varkappa^2 p^2\,(\gamma/2)},$$

$$V^2\,(x_0) = \frac{\left\{\left[2\varkappa + \varkappa p\,(\gamma/2)\frac{x_0}{N}\frac{dN}{dx_0}\right] r_0 - 2\varkappa r\,(\gamma/2)\right\}^2 x_0^2 N^4\,(x_0)}{\left[2\varkappa + \varkappa p\,(\gamma/2)\frac{x_0}{N}\frac{dN}{dx_0}\right]^2 + 2}. \qquad (I.55)$$

We perceive at once that the expression for $\mathscr{E}^2$ is very reminiscent of the expression for P^2. The one difference is that now instead of $M(x)$ we have $L(x_0)$, and instead of $r^2 x_0^2 N^4(x_0)$ we have $V^2(x_0)$. Consequently, the optimum conditions for width measurements can be distinguished from the optimum conditions for optical density measurement only in their dependence on x_0.

Proceeding as in the analysis of P^2, we again obtain

$$y_m = \frac{z_m}{2}\sqrt{\frac{a}{b}}$$

with the same relations between the different types of error as in Eqs. (I.33) and (I.34).

The expressions for z_m and $\mathscr{E}_m^2$ derive automatically from (I.35) and (I.36) by substituting L for M and V^2 for $x_0^2 r^2 N^4$. This yields

$$z_m = \left[\frac{8U}{5a^2 V^2\,(x_0)\,\sqrt{a/b}}\right]^{1/5}, \qquad (I.56)$$

$$\mathscr{E}_m^2 = L\,(x_0) 2^{-9/5} 5^{1/5} 3^2\, a^{1/5} b^{1/5} V^{16/5}(x_0)\, U^{4/5}. \qquad (I.57)$$

The relation $\mathscr{E}_m$ is depicted in Fig. 5. Curve a refers to a band of dispersion configuration; for the solid curve the function $N(x_0)$ has been calculated from Eq. (I.43) assuming a dispersion configuration for $D(\lambda)$, for the dashed curve it has been calculated from Eq. (I.44) assuming a dispersion configuration for $x(\lambda)$. Curve b

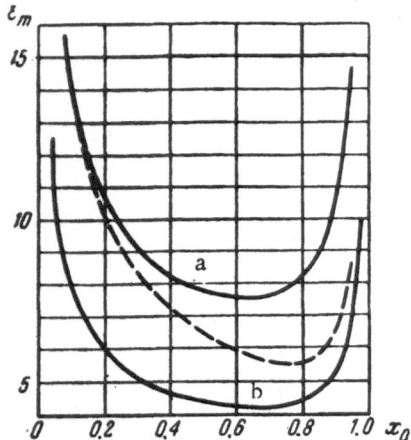

Fig. 5. Dependence of the optimum relative error in measuring the half-width γ_D of the optical density band on the peak absorption.

refers to a Gaussian configuration for $D(\lambda)$; $N(x_0)$ has been calculated from Eq. (I.45). A comparison of Figs. 5 and 3 discloses the existence of an interval of x_0 in which the errors in measuring D_0 and γ_D simultaneously approach a minimum. This interval falls approximately within the limits $0.3 < x_0 < 0.85$.

All that is left to clear up at this point is how to obtain the z_m values for measurement of γ_D. To do this we need to compare $V^2(x_0)$ and $r^2 x_0^2 N^4(x_0)$. Calculations have shown that the difference is small for values of x_0 lying within the pertinent interval. Since these quantities occur to the $\frac{1}{9}$ power, they may be disregarded.

Consequently, the optimum conditions for measuring D_0 and γ_D may be treated as roughly congruent.

Some specific examples are in order.

Let the instrument function be determined only by the slits, let $S_1 = S_2 = S$, and let the frequency characteristic of the registration system be determined by an RC circuit with time constant τ. In this case $\theta = \tau v$ is the total shift of the contour as a whole,

$$a = 1/12, \qquad b = 1/2, \qquad \tau_m v = \frac{S_m}{2} \sqrt{1/6} \simeq 0.204\, S_m. \tag{I.58}$$

If we assume that the form of the instrument function is almost Gaussian with a width S, then

$$\tau_m v = 0.212\, S_m. \tag{I.59}$$

We see that in both cases $\tau_m v \simeq \frac{1}{5} S_m$.

Consider now two distinct lines of different configuration (Gaussian and dispersion); we will investigate measurements of γ_D and the intensity at the peak of the lines, where φ'' is a maximum and, consequently, the error measurement will be a maximum. We will carry out our calculations for $x_0 = 0.6$, which corresponds roughly to the error minimum.

For the line of Gaussian configuration

$$r_0 = -8 \ln 2, \qquad N(0.6) = 0.875.$$

For the line of dispersion configuration

$$r_0 = -8, \qquad N(0.6) = 0.798.$$

From (I.35), (I.36), and (I.56), (I.57) we obtain for the Gaussian band

$$
\begin{aligned}
z_{mD} &= 1.66\, U^{1/9}, & P_m &= 2.55\, U^{2/9}, \\
z_{m\gamma} &= 1.59\, U^{1/9}, & \mathscr{E}_m &= 2.42\, U^{2/9},
\end{aligned}
\tag{I.60}
$$

and for the dispersion band

$$
\begin{aligned}
z_{mD} &= 1.58\, U^{1/9}, & P_m &= 2.90\, U^{2/9}, \\
z_{m\gamma} &= 1.69\, U^{1/9}, & \mathscr{E}_m &= 4.34\, U^{2/9}.
\end{aligned}
\tag{I.61}
$$

Since the shape of individual absorption bands clearly varies somewhere between dispersion and Gaussian configurations, the resultant equations permit a coarse estimate of the optimum conditions for measuring the peak optical intensity and width of the band for almost all bands encountered in practice. The fluctuation of shape is manifested only in the coefficients of Eqs. (I.60) and (I.61), and not very severely at that.

Let us compare our optimum measurement conditions with the recommendations of other authors. There is a large body of papers in which the distorting influence of various instrument factors is analyzed and any of a number of recommendations are given for the choice of suitable conditions of measurement. Thus, for example, in [49, 50] the effect of the recording intrumentation response in the scanning of a spectrum in considered and the conditions are described under which the associated error will not exceed a preset value. The influence of the slits is investigated in [51-53]. In the cycle of investigations [54-56] all of the instrumental distortion factors are examined: slit effects, the time constant, noise level, etc. However, the recommendationa as to choice of measurement conditions are derived from the requirement that a prescribed **signal-to-noise ratio** be maintained, essentially without regard for the systematic distortions associated with the slits.

The authors of [57, 58] also rely on the requirement of preserving a specific signal-to-noise ratio. All of these papers fail to solve the problem of finding optimum measurement conditions. Those which neglect to consider all of the error sources simultaneously, of course, do not expose optimum measurement conditions, since the different types of error are interrelated, and it is always possible to diminish errors of one type while increasing exorbitantly errors of the other type. When this happens, of course, the total error may be quite far removed from the optimum. Consequently, it is impossible, if any source of error is ignored, to obtain the optimum conditions for measurement. It is not at all difficult to see that the requirement of a constant predetermined signal-to-noise ratio does not imply minimum total error. In fact, if we specify some definite signal-to-noise ratio the systematic error will predominate for narrow bands, while, on the other hand, for wide bands the systematic error may prove to be negligible. In both instances the optimum may be approached by changing the ratio between the random and systematic errors in the first case toward lower systematic and higher random error, doing just the opposite in the second case (for example, by varying the slit width and time constant).

Hence the disparity between our own results as to the choice of experimental conditions and the results of the abovementioned papers follows a pattern of regularity. In specific cases the difference in recommendations regarding the assignment of certain parameters can prove to be considerable. For instance, from the requirement of a minimum total mean square error we obtained the condition $\tau_m v \sim S_m$, whereas the above papers give the relation $\tau \sim 1/S^4$ on the basis of maintaining a prescribed signal-to-noise ratio.

In [38, 39] the optimum conditions are derived as in our case, by insisting upon a minimum total error. The equations obtained in those papers bear a resemblance to our own (in particular they give $\tau_m v \sim S_m$). However, as we already noted, an approximation is used in which only the first-order error is described, i. e., essentially the shift error, which we did not take into account. These results cannot, therefore, be regarded as comparable with ours.

As for the choice of optimum absorption (or optimum layer thickness, or optimum optical density), our recommendations are not too inconsistent with the well-known results of [28-30]. A certain displacement of the optimum interval of x_0 is related primarily to the fact that in our case the analysis was made with a stipulated choice of optimum values for other parameters (S, τ) and with allowance for the variation in width of the absorption band as a function of the absorption at peak value.

CHAPTER II

REDUCTION TO AN IDEAL INSTRUMENT [31, 59, 60]

§1. Influence of Reduction on the Accuracy of Measurements

In the preceding chapter we investigated the optimum conditions for working without reduction and explained what kind of measurement accuracy could be obtained in this mode of operation. The following question logically arises at this point: what advantages are offered by the application of reduction methods. To find the answer we must evaluate the accuracy attainable with the use of a particular reduction method, and we must decide how much advantage in accuracy is gained at the expense of effort expended in actually performing the reduction. We first ask the question, is it possible in general by the application of reduction to really gain in accuracy? Reduction is usually treated by itself, without regard for the transformation of error. This has created the impression that reduction increases measurement accuracy without fail. The arguments developed in the Introduction show that any reduction increases the role of the random error, rendering the answer to our question not immediately obvious. We will look into this question in detail.

We will consider the transmission of a test signal through a spectral instrument and analyze the origin of the various distortions by stages. The investigated spectrum first passes the monochromator. Here it is subjected to certain systematic distortions associated with the instrument function of the monochromator. If we are concerned with infrared spectroscopy there is no random error at this stage. It is known [61-65] that this type of error is associated primarily with noise in the radiation receiver (thermocouple, bolometer, etc.), i.e., it arises at the input to the registration system.

The values and statistical properties of this noise at the output of the total instrumentation are determined by the gain and frequency characteristic of the amplifying section. The amplitude of the light signal at the output of the total instrumentation, however, is determined by the luminous intensity of the light passing through the monochromator, sensitivity of the receiver, and gain of the amplifier. Additional systematic distortions occur in the registration system in connection with its response, their nature and extent being governed by the frequency characteristic $K(\omega)$.

It is important to recognize the fact that the systematic distortions due to the monochromator instrument function appear before the onset of noise, while the response type distortions appear afterwards. This distinction is important and means that the reduction of these two types of systematic distortion will produce different results.

In order to clafify this assertion we will analyze separately the reduction of systematic distortion introduced by the monochromator and the reduction of response type distortion originating in the registration system. The Fourier transform of the distribution $h(\lambda)$ measured at the instrument output, as we saw already, has the form

$$H(\omega) = \Phi(\omega) A(\omega) K(\omega) + X(\omega). \tag{II.1}$$

If the system is linear the relative random error (ratio of light signal-to-noise) does not depend on the gain. We will assume that the distortion in the monochromator is small, $A(\omega) \simeq 1$, and inspect the result of reduction of the distortions that arise in the registration system. As we saw in the Introduction, reduction means multiplying the function H(ω) by the function G(ω). The function G(ω) is chose such that it will approximate $1/K(\omega)$ in some interval of ω. The product $K(\omega) G(\omega)$ describes the residual frequency characteristic of the registration system. After carrying out the reduction process we have

$$\Phi_a(\omega) = \Phi(\omega) K(\omega) G(\omega), \tag{II.2}$$

$$\overline{\Delta\varphi^2} = \frac{1}{\pi} \int\limits_0^\infty \Psi(\omega) |K(\omega) G(\omega)|^2 d\omega. \tag{II.3}$$

As always, reduction decreases the systematic error and increases the random error.

As a result we obtain at the output that distribution $\varphi_a(\lambda)$ and noise which we would have if the measurements were made with a registration system having the frequency characteristic $K(\omega) G(\omega)$. Consequently, reduction in this case acts simply as if to change the frequency characteristic of the system. This is related to the fact that $K(\omega)$ operates identically on both the noise and the light signal. Inasmuch as it is usually a simple matter to change the frequency characteristic of the system (for example, by changing its time constant or the parameters of the amplifying section), under normal infrared spectroscopy conditions there is no point in analytically eliminating the systematic error of the registration system. It is simpler to set the necessary $K(\omega)$ beforehand, prior to recording the spectrum. Specifically, if the light receiver itself is too slow in response to provide a sufficiently broad transmission band, subsequent correction can be incorporated into the amplifying section to obtain the required $K(\omega)$ [66-68].

Let us now investigate the reduction of systematic distortions borne by the monochromator. Unlike the frequency characteristic of the registration system, the instrument function of the monochromator affects the signal even before the onset of noise and, hence, has no influence on the latter. What happens if we increase the width $a(\lambda)$? The systematic error will increase and the noise will not in general be affected. However, broadening of the instrument function usually involves a rapid increase in the intensity of the light transmitted through the instrumentation, so that the amplitude of the light signal increases in proportion to the noise. As long as the instrument function is narrow relative to the measured structure, the systematic error associated with it will be small (less than the noise) and it will prove advantageous to broaden it. But at the instant that the systematic error begins to provide the major contribution any further broadening of $a(\lambda)$ becomes disadvantageous. As shown in Chapter I, when working without reduction the accuracy is a maximum when the systematic error is approximately equal to the random error. This, however, does not exhaust the possibilities for reducing the error. Let us look back at Fig. 1, in which the form of the functions $\Phi(\omega)$, $A(\omega)$, $F(\omega)$, and $\Psi(\omega)$ is illustrated [we should in general consider complex functions, but for simplicity we are presenting here the case when they are real, which corresponds to symmetrical functions $\varphi(\lambda)$, $a(\lambda)$, and $f(\lambda)$]. We agree that we are going to increase the width of the instrument function beyond that optimum value which corresponds to working without reduction. Then the function $A(\omega)$, and hence $F(\omega)$, becomes still narrower, but the value of $F(\omega)$ increases due to the greater amount of light transmitted through the instrument. If we assume that the contours of $\varphi(\lambda)$, $a(\lambda)$, and $f(\lambda)$ are symmetrical and have maxima at $\lambda = 0$ (actually the variable $\lambda - \lambda_0$ is used) the area under the curve $F(\omega)$ will be equal to the amplitude of the observed contour at the maximum:

$$f_0 = \frac{1}{2\pi} \int\limits_{-\infty}^\infty F(\omega) d\omega.$$

Constricting $F(\omega)$ causes f_0 to become smaller than if instrument distortions were not present. The function $\Psi(\omega)$, on the other hand, is not affected; the noise maintains its same level [as opposed to this, $K(\omega)$ operates both on $\Phi(\omega)$ and on $\Psi(\omega)$].

We now apply reduction, i. e., we multiply $F(\omega)$ by $G(\omega)$ (see Fig. 1), where $G(\omega)$ approximates $A^{-1}(\omega)$ to the extent that it is almost congruent with $A^{-1}(\omega)$ for small ω and falls below it for large ω, delineating the region of "reducible" frequencies. After reduction we have the residual instrument function with Fourier transform $A(\omega) G(\omega)$, which may be imparted any width desired and therefore reduces the systematic error to a certain extent. The penalty for this is increased noise, since the reduction affects the entire output signal. The magnitude of the noise after reduction becomes

$$\overline{\Delta \varphi_0^2} = \frac{1}{\pi} \int_0^\infty \Psi(\omega) |G(\omega)|^2 d\omega. \qquad (II.4)$$

Consequently, the application of reduction in this case, as opposed to the reduction of response type distortions, is not equivalent to a simple translation to a narrower instrument function. This means that it is possible in principle to enhance the accuracy by mathematical processing of the output contour with an appropriate choice of conditions for recording of the spectrum. In doing this we essentially build in an increased systematic error, thereby gaining in luminosity of the instrument; then we analytically decrease the systematic error by increasing the random error. Whether we win accuracy as the result of this operation depends on the extent to which the systematic error and luminosity of the instrument increase when the instrument function is broadened. Consequently, this depends on the relationship between the width of the instrument function and the luminous flux transmitted through the instrument, as well as on the width and shape of the investigated structure and instrument function, which are factors governing the systematic error. Moreover, this depends on the attributes of the measurement error. Hence the conceptual feasibility of increasing the measurement accuracy by reduction exists, but the gain in accuracy and the conditions under which it can be obtained depend on the special characteristics of the instrument and investigated object; they must be subjected to ad hoc calculations with allowance for the dependences indicated.

Let us consider the mean square total measurement error $\varphi(\lambda)$ after reduction. Making use of (II.2) and (II.3), we find

$$\overline{\Delta \varphi^2(\lambda)} = \frac{1}{\pi} \int_0^\infty \Psi(\omega) |K(\omega) G(\omega)|^2 d\omega + [\varphi(\lambda) - h(\lambda)]^2$$

$$= \frac{1}{\pi} \int_0^\infty \Psi(\omega) |K(\omega) G(\omega)|^2 d\omega + \left[\frac{1}{2\pi} \int_{-\infty}^\infty \Phi(\omega)(1 - K(\omega) A(\omega) G(\omega)) e^{-i\omega\lambda} d\omega \right]^2. \qquad (II.5)$$

The first term in this equation describes the random error, transformed as the result of reduction; the second term represents the residual systematic error. $G(\omega)$ is a function approximating $A^{-1}(\omega)$ in some frequency interval. The problem is now contained in varying the conditions of measurement [the parameters determining $A(\omega)$ and $K(\omega)$] and the function $G(\omega)$ (reduction conditions) in order to ascertain the conditions that will provide minimum error in the final result. The general solution of this problem is very complex. We will carry it through for a special case with appreciable simplifications.

§2. Optimum Measurement Conditions and Optimum Reduction by the Fourier-Transform Method

For computational simplicity we will not consider in this section the absorption dependence of the error. As shown by a general analysis conducted by S. G. Rautian [20, 32] on the peak-absorption dependence of the measurement error, an optimum value of x_0 for all reduction methods can be chosen within rather broad limits. Consequently, it is possible, without a great sacrifice in accuracy, to preassign a definite x_0 and then to operate with the absorption contour. The width of this contour will be denoted by γ.

We have seen that the reduction of response type distortions is pointless. On the other hand, it was shown in Chapter I that it is advantageous to choose $K(\omega)$ such that the systematic error due to scanning will be small in comparison with the error due to the instrument function. This implies that $K(\omega)$ must be considerably broader than $A(\omega)$ and $\Phi(\omega)$. Without incurring large error we can let $K(\omega) = 1$, i.e., in general ignore the response distortions due to scanning. This yields

$$\overline{\Delta\varphi^2(\lambda)} = \frac{1}{\pi}\int\limits_0^\infty \Psi(\omega)\,|G(\omega)|^2\,d\omega + \left|\frac{1}{2\pi}\int\limits_{-\infty}^\infty \Phi(\omega)\,[1 - A(\omega)\,G(\omega)]\,e^{-i\omega\lambda}\,d\omega\right|^2. \qquad (\text{II.6})$$

It is necessary now to find those conditions of measurement and function $G(\omega)$ such that Eq. (II.6) will revert to a minimum. This is an extremely complex variational problem. Rather than solve it, we will assume a predetermined simple form for $G(\omega)$, thus reducing the problem to the ordinary minimization of a function of several parameters.

Consider the following form of $G(\omega)$:

$$G(\omega) = \begin{cases} A^{-1}(\omega) & \omega \leq \omega_0, \\ 0 & \omega > \omega_0. \end{cases} \qquad (\text{II.7})$$

$G(\omega)$ precisely reproduces $A^{-1}(\omega)$ in some frequency interval, then veers sharply away from it. Investigating for simplicity the symmetrical functions $\varphi(\lambda)$ and $a(\lambda)$, we find

$$\overline{\Delta\varphi^2(\lambda)} = \frac{1}{\pi}\int\limits_0^{\omega_0} \Psi(\omega)\,A^{-2}(\omega)\,d\omega + \left[\frac{1}{\pi}\int\limits_{\omega_0}^\infty \Phi(\omega)\,e^{-i\omega\lambda}\,d\omega\right]^2. \qquad (\text{II.8})$$

For the band peak $\lambda = 0$ we obtain

$$\overline{\Delta\varphi_0^2} = \frac{1}{\pi}\int\limits_0^{\omega_0} \frac{\Psi(\omega)}{A^2(\omega)}\,d\omega + \frac{1}{\pi^2}\left|\int\limits_{\omega_0}^\infty \Phi(\omega)\,d\omega\right|^2. \qquad (\text{II.9})$$

The problem of finding an optimum reduction amounts in our case to finding the optimum cutoff frequency ω_0 at which (II.9) becomes a minimum. The minimal condition

$$\frac{\partial}{\partial\omega_0}\overline{(\Delta\varphi_0^2)} = \frac{1}{\pi}\frac{\Psi(\omega_0)}{A^2(\omega_0)} - \frac{2}{\pi^2}\Phi(\omega_0)\int\limits_{\omega_0}^\infty \Phi(\omega)\,d\omega = 0 \qquad (\text{II.10})$$

yields

$$\Psi(\omega_0) = \frac{2}{\pi}\Phi(\omega_0)\,A^2(\omega_0)\int\limits_{\omega_0}^\infty \Phi(\omega)\,d\omega. \qquad (\text{II.11})$$

Our task is to determine ω_0 from this equation. To do this we need to know $F(\omega)$, $A(\omega)$, $\psi(\omega)$ and be able to evaluate the integral

$$\int\limits_{\omega_0}^\infty \Phi(\omega)\,d\omega = \int\limits_{\omega_0}^\infty \frac{F(\omega)}{A(\omega)}\,d\omega.$$

In the event that the form of $\varphi(\lambda)$ is not known beforehand, this integral can only be evaluated by computing $F(\omega)$ and $A(\omega)$ in a sufficiently broad interval for $\omega > \omega_0$, i. e., in that interval which is to be neglected below and not used for calculating φ_0. Furthermore, we will be able to evaluate this integral only with an appreciable error, since the quantity $\Phi(\omega)$ in this interval of ω will be small in comparison with the noise. In principle this induces additional error in φ_0, which will have to be evaluated.

If the form of $\varphi(\lambda)$ is known beforehand the integral in (II.11) and the optimum cutoff frequency ω_0 can be expressed in terms of the parameters of the band being measured.

Orienting ourselves once again in the direction of typical conditions for infrared spectroscopy, we will adopt the following assumptions relative to the form of the functions pertinent to our discussion.

1. An instrument function of Gaussian form whose width S is equal to the slit width is

$$A(\omega) = e^{-\frac{\omega^2 S^2}{16 \ln 2}}. \tag{II.12}$$

2. The true dispersion configuration is

$$\Phi(\omega) = \varphi_0 \pi \gamma e^{-|\omega| \frac{\gamma}{2}}, \qquad \varphi(\lambda) = \frac{\varphi_0}{1 + 4\lambda^2/\gamma^2}. \tag{II.13}$$

3. We will assume that the spectrum of the receiver noise is uniform; for thermal receivers, such as thermocouples, bolometers, etc., this is normally the case [63, 65]. We will consider the output noise spectrum to be determined by the frequency characteristic of an RC circuit with time constant τ, the latter being sufficiently small that the response type distortions are negligible. We can assume that in the ω interval of interest*

$$K(\omega) = \frac{1}{1 + i\omega\tau v} \simeq 1, \quad \Psi(\omega) = \Psi_0 = \text{const.} \tag{II.14}$$

The connection between ψ_0 and the noise type error is given by the expression

$$\Psi_0 = 2\tau v \overline{\Delta l^2}. \tag{II.15}$$

Letting $\Delta l^2/I_0^2 = E/\tau S^4$, we obtain

$$\Psi_0 = 2 \frac{I_0^2 E v}{S^4}. \tag{II.16}$$

Using our assumptions, we now obtain instead of (II.9)

$$\overline{\Delta\varphi_0^2} = \frac{1}{\pi} \int_0^{\omega_0} \Psi_0 e^{\frac{\omega^2 S^2}{8 \ln 2}} d\omega + \left[\frac{1}{\pi} \int_{\omega_0}^{\infty} \varphi_0 \pi \gamma e^{-\omega \frac{\gamma}{2}} d\omega \right]^2, \tag{II.17}$$

whence

$$\frac{\overline{\Delta\varphi_0^2}}{\varphi_0^2} = 4e^{-\omega_0 \gamma} + \frac{2I_0}{\pi\varphi_0^2} \frac{Ev}{S^4} \int_0^{\omega_0} e^{\frac{\omega^2 S^2}{8 \ln 2}} d\omega. \tag{II.18}$$

As usual, in this equation the first term describes the residual systematic error, the second is the noise transformed by reduction.

To determine the optimum conditions for measurement and reduction we need now to find the minimum of (II.18) with respect to S and ω_0. Following standard procedure, we obtain the system of equations

$$\omega_0 = \frac{4 \ln 2}{S} \left(\sqrt{\frac{\gamma^2}{S^2} - \frac{1}{2 \ln 2} \ln \frac{I_0^2 Ev}{2\pi\varphi_0^2 \gamma S^4}} - \frac{\gamma}{S} \right), \quad \int_0^{\omega_0} e^{\frac{\omega^2 S^2}{8 \ln 2}} d\omega - \frac{S^2}{16 \ln 2} \int_0^{\omega_0} \omega^2 e^{\frac{\omega^2 S^2}{8 \ln 2}} d\omega = 0. \tag{II.19}$$

---

*To avoid misunderstanding we recall that ω is not the usual circular frequency with dimensionality $1/t$, where t is the time, but has the dimensions of $1/\lambda = 1/vt$. In correspondence with this the functions $K(\omega)$ and $\psi(\omega)$ include dependence on the scan rate v.

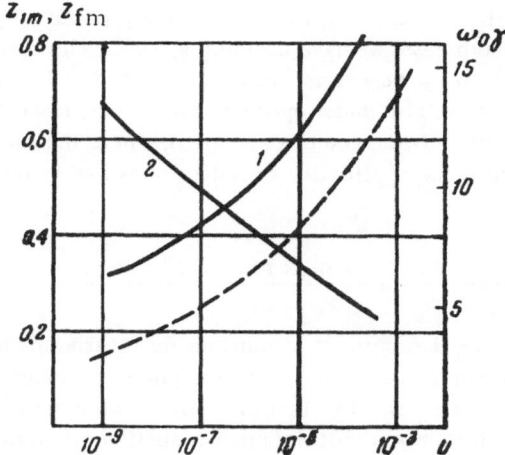

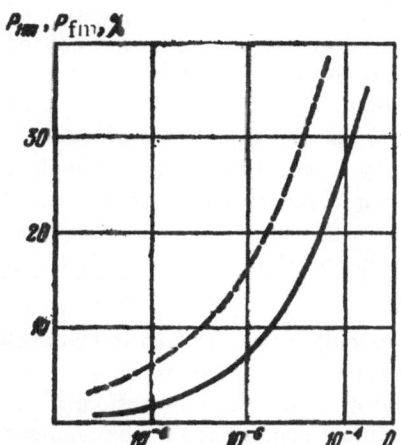

Fig. 6. Dependence of the optimum values of z_{fm} (curve 1) and $\omega_0\lambda$ (curve 2) on the parameter U for the method of reduction by Fourier transforms. The dashed curve represents the optimum values of z_{1m} for the case without reduction.

Fig. 7. Dependence on U of the total optimum error in measuring optical density with reduction by Fourier transforms. The dashed curve represents the total optimum error for the case without reduction.

The first of these equations determines the optimum cutoff frequency ω_0 of the spectrum in terms of the slit width S.

It is impractical to solve this system. We instead carry out a direct calculation of $\overline{\Delta D^2}/D^2$ using the relation that determines the optimum ω_0 and the following equations:

$$P_f^2 = \frac{\overline{\Delta D^2}}{D^2} = M(x_0)\frac{\overline{\Delta\varphi_0^2}}{\varphi_0^2} = M(x_0)\left\{8\sqrt{2\ln 2}\,\frac{K}{z^5}\int_0^{\frac{wz}{\sqrt{8\ln 2}}} e^{t^2}\,dt + 4e^{-w}\right\},\qquad(\text{II}.20)$$

$$w = \omega_0\gamma = \frac{4\ln 2}{z}\left(\sqrt{\frac{1}{z^2} - \frac{1}{2\ln 2}\ln\frac{Q}{z^4}} - \frac{1}{z}\right),\qquad(\text{II}.21)$$

where we have introduced the dimensionless parameters

$$z = \frac{S}{\gamma}\,;\quad w = \omega_0\gamma,\quad Q = \frac{I_0^2 Ev}{2\pi\varphi_0^2\gamma^5} = \frac{U}{2\pi x_0^2}\,,\qquad(\text{II}.22)$$

and the factor $M(x_0)$ has the form

$$M(x_0) = \left[\frac{x_0}{(1 - x_0)\ln(1 - x_0)}\right]^2.$$

The results of the calculation are shown in Fig. 6, which illustrates the relations $z_{fm}(U)$ and $\omega_{0m}\gamma(U)$, and in Fig. 7, which illustrates the relation $P_{fm}(U)$.* We note also that the product $\omega_0 S_m$ remains almost constant in

---

*The subscript "m," as usual, denotes the optimum value of the indicated variable; the subscript "f" signifies that the given variable is obtained by the Fourier transform method of reduction.

the interval $10^{-9} < U < 10^{-4}$ and approximately equal to 4.2. Inasmuch as we have not taken into account the dependence of the width and configuration of the absorption band on x_0, the curves for $z_{1m}(U)$ and $P_{1m}(U)$ are also calculated without regard for this dependence. In both cases we have assumed $x_0 = 0.5$. The ratio P_{1m}/P_{fm}, which determines the increase in accuracy due to reduction by the Fourier transform method under optimum conditions relative to simple measurement without reduction, also under optimum conditions, is shown in Fig. 8. It is apparent that the application of reduction is especially advantageous for small U, where the increase in accuracy may be as much as four- or fivefold. For large U the application of reduction is barely justifiable, since the gain in accuracy is comparatively slight.

§3. Optimum Conditions with Exact Reduction. Comparison of Various Reduction Methods.

The calculations performed in §2 enable us to choose the optimum conditions for measurement with reduction by the Fourier transform method. Used in broad application, this method requires computer techniques for computation of the Fourier transforms, which is not always possible. It is of interest, therefore, to try and find the optimum conditions for measurement and reduction for some of the other reduction methods in use. The calculations for the Rayleigh method were made by S. G. Rautian [20, 59]. We will consider one of the exact methods of reduction, the method of Willis [69]. It is based on the assumption of a Gaussian configuration on the part of both the measured band and the instrument function, and was decided upon for the estimation of optimum conditions in view of the relative simplicity of its calculations.

In examining this method we must realize the role of the measurement error. It was asserted in the Introduction that in general exact reduction is unfeasible because of the impossibility of reducing $\Phi(\omega)$ over the entire range of ω. However, the indicated method does impart a definite form to $\Phi(\omega)$, so that by effectively limiting the interval of ω it becomes reducible. Knowing the instrument function and shape of the measured distribution, we can determine the parameters of the contour being measured from the parameters of the observed contour. Consequently, the result obtained will not contain systematic error. But the parameters of the observed contour are known to have a certain error originating from the random error of measurement. As a result of this the computed parameters of the measured contour will also contain error. We are concerned, therefore, with estimating the magnitude of the error in the final result as a function of the conditions of reduction and conditions of measurement, and selecting these conditions such that the error in the final result will be minimized.

We will perform the necessary analysis with certain simplifications. Above all, we will disregard the dependence of the configuration and width of the absorption band on the absorption at its peak value. We will assume that the shape of the absorption band is Gaussian and that its width γ at the half-power points is

$$\varphi(\lambda) = \varphi_0 e^{-\frac{4\ln 2}{\gamma^2}\lambda^2}. \qquad (II.23)$$

The instrument function is also assumed to have a Gaussian configuration:

$$a(\lambda) = \frac{2}{\alpha}\sqrt{\frac{\ln 2}{\pi}}\, e^{-\frac{4\ln 2}{\alpha^2}\lambda^2}. \qquad (II.24)$$

Here $\alpha = S + c$ is the width of the instrument function at the half-power points, S is the slit width of the monochromator (the slits are assumed equal), c is the nonslit correction to the width of the instrument function [20, 32].

The observed band is in this case described by the expression

$$f(\lambda) = f_0 e^{-\frac{4\ln 2}{\delta^2}\lambda^2},$$
$$f_0 = \gamma\varphi_0/\delta, \qquad \delta^2 = \gamma^2 + \alpha^2. \qquad (II.25)$$

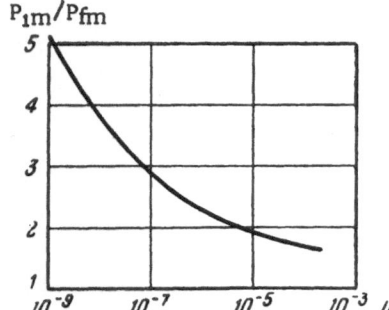

Fig. 8. Dependence on U of the optimum error ratio in measuring the optical density, P_{1m}/P_{fm}.

With these equations the parameters ϕ_0 and γ of the true contour are readily calculated in terms of δ and f from the observed contour.

Since the method uses exact reduction, it would be natural to expect the optimum values of the slits to be relatively large. The random error in measuring the observed contour under optimum conditions may prove to be very small, in fact comparable with the "mechanical" instrument error. It so happens that by increasing the slit width we can only reduce the measurement error to some value defined by the class of precision of the recording arrangement. Any further broadening of the slits would be pointless. The results of measurements of the appropriate accuracy characteristics of a typical instrument are given in the Conclusion. It turns out that if one is going to perform an analysis by the Willis method it is necessary to account for the mechanical error. In accordance with the result obtained in [31], therefore, we will adopt the following approximate error characteristic:

$$\frac{\overline{\Delta l^2}}{l_0^2} \begin{cases} \dfrac{E}{\tau S^4} = \dfrac{A^2}{S^4} & S^2 \leqslant \dfrac{A}{B}, \\[2ex] B^2 & S^2 > \dfrac{A}{B}. \end{cases} \qquad (\text{II.26})$$

As in §2, we will not take into account the response type distortions in the registration system. We will consider that τv is fixed and sufficiently small. For $y = \tau v/\gamma$ we will assume a value $y = 1/10$.

Omitting the cumbersome computations (for more details see [59]), which are similar to the usual error calculations, we will cite the expression for the relative mean square error in measuring the optical density at the band peak:

$$P_3^2 = \frac{\overline{\Delta D^2}}{D^2} = M(x_0) \left\{ \left[1 + \left(\frac{1-x_0}{x_0} \right)^2 \right] B^2 + \frac{U}{y x_0^2 (z-g)} R(z,n) \right\},$$

$$(\text{II.27})$$

$$R(z,n) = (1+z^2) \left\{ \frac{1}{2} \left[1 + \frac{W}{2n} z^2 - \frac{1}{\sqrt{1+z^3}} \right]^2 + \left[1 + \frac{W}{2n} z^2 \right]^2 + \frac{W^2}{8} z^4 \right\}.$$

It is assumed here that

$$z = \frac{\alpha}{\gamma} = \frac{S+c}{\gamma}, \qquad g = \frac{c}{\gamma}, \qquad U = \frac{Ev}{\gamma^5}, \qquad W = \frac{n^{\ln n}}{\ln n}.$$

The measured width of the band is arbitrarily taken at the height f_0/n. In this formula the factor

$$M(x_0) = \left[\frac{x_0}{(1-x_0) \ln (1-x_0)} \right]^2$$

stems from translating the absorption error to the optical density error. The first term in the braces is elicited by the error in measuring the lines of total transmission and total absorption, which are assumed to be recorded with minimum error B. The second term describes the random error of the final result, where the factor $R(z,n)$ represents the increase in random error with reduction and the expression $U/y x_0^2 (z-g)^4$ corresponds to the random error in the observed contour. For $S^2 > A/B$ (wide slits) this expression is replaced by B according to Eq. (II.26).

To find the optimum conditions we need now to minimize this expression with respect to three parameters: x_0, z, n. The optimum value of the peak absorption [32] will be $x_{0sm} = 0.632$. Analysis of the function $UR(z,n)/(z-g)^4$, the properties of which influence the optimum values of n and z, yields the following conclusions. For any values of U and g the optimum value of n will lie between the limits $2.4 < n < e = 2.72$, i.e., essentially one value can be used in every case. We took the value $n = 2.72$ and computed z_{3m} and P_{3m}. The results are shown in Fig. 9. The kink in the curve at $U = 1.6 \cdot 10^{-6}$ means that for small U mechanical error is involved. As evident from Fig. 9, with exact reduction the spectrum will be recorded in many cases

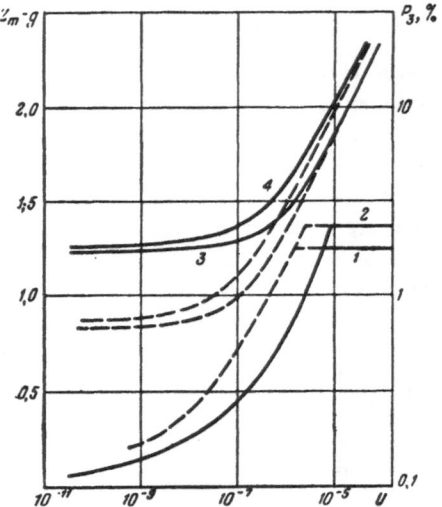

Fig. 9. Dependence of the optimum slit width $S_{3m}/\gamma = z_{3m} - g$ and optimum error P_{3m} (exact reduction by Willis method) on the parameter U. 1, 2) $z_{3m} - g$ for $g = 0$ and 0.2, $B = 0.5\%$; 3, 4) P_{3m} for $g = 0$ and 0.2, $B = 0.5\%$; the dashed curves are for $B = 0.2\%$.

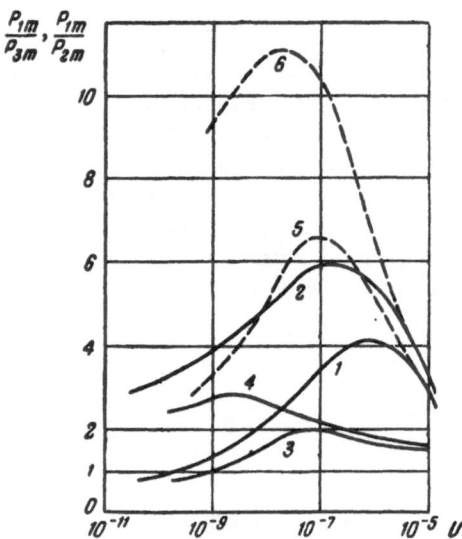

Fig. 10. Ratios P_{1m}/P_{3m} and P_{1m}/P_{2m} of optimum errors in measuring optical density, as a function of U. 1, 2) P_{1m}/P_{3m} for $g = 0$ and 0.2, $B = 0.5\%$; 3, 4) P_{1m}/P_{2m} for $g = 0$ and 0.2, $B = 0.5\%$; 5, 6) P_{1m}/P_{3m} for $g = 0$ and 0.2, $B = 0.2\%$.

with slit widths determined by the condition $S^2 = A/B$, and only for $U > 6.4 \cdot 10^{-6}$ will receiver noise be significant. As expected, the optimum slit widths turn out to be even larger than with reduction by the Rayleigh method. Agreement is observed only in those cases when reduction by the Rayleigh method also involves mechanical error in the recording of the spectrum. We not that this agreement of conditions will occur for any reduction method and any spectral distribution of energy. We notice a shift of the kink in the curves $z_{3m} - g$ and an abrupt decrease in P_{3m} wherever B plays a major role.

For a comparison of the various reduction methods from the point of view of attainable accuracy it seems reasonable to take the ratios of the optical density errors obtained under conditions which are optimal for each method, i.e., P_{1m}/P_{3m} and P_{1m}/P_{2m}, which demonstrate what accuracy is gained in going from the case without reduction to exact reduction and to reduction by the Rayleigh method.* These ratios are shown in Fig. 10 as a function of U.

As evident from Fig. 10, the gain in accuracy depends on the quantity U, i.e., on the precision characteristics of the instrument and width of the band, and on the method of reduction, the gain being greatest for exact reduction. It is important to realize that the gain in accuracy depends very strongly on the mechanical error. For $B = 0.5\%$ the accuracy is increased sixfold with exact reduction, and threefold in the case of Rayleigh reduction, whereas for $B = 0.2\%$ the ratio P_{1m}/P_{3m} may be as high as 11. Consequently, only with small mechanical error does exact reduction offer any essential advantage. This result also indicates how important it is to know the magnitude of B in order to give a justifiable evaluation of the suitability of reduction and of the attainable accuracy in determining spectroscopic vairables. Despite this indication, however, publications on the subject generally do not recognize this important characteristic of instrumentation. We note that modern instruments, as a rule, have a registration system in the precision class 0.5, i.e., $B = 0.5\%$. For efficient utilization of reduction one must seriously inquire into the possibility of improving this characteristic of the instrumentation.

---

*The Rayleigh method calculations have been made by S. G. Rautian [29, 59]. We use these results here, denoting the mean square error in measuring the optical density with Rayleigh reduction under optimum conditions by the symbol P_{2m}.

CHAPTER III

RESOLVING POWER
OF SPECTRAL INSTRUMENTS [31, 35, 70, 71]

The problem of the resolving power of spectral instruments constitutes part of the more general problem of the investigation of and correction for instrument distortions induced by the equipment into the investigated distribution. There are two basic features which distinguish from the overall problem area that category generally known as the "resolution problem" in spectral apparatus. The first feature is that the instrument distortion is studied as to its effects on a spectrum of fairly complex structure, at the very simplest on two identical monochromatic lines spaced very closely together. The second feature is that, instead of investigating the accuracy of some measured characteristic of the spectrum to be studied, it is customary to consider the possibility of discovering purely qualitatively a particular structure of the spectrum. In the simplest case one may wish to disclose two lines.

It is important to realize that until the present time no universally accepted approach has been established for this entire problem category. The fundamental concept in particular—resolving power of the spectral instrument—is treated differently by different authors. It is useful, in this respect, to summarize briefly the main trends in the development of research on the resolution problem.

The first to inquire into the limit of resolution was clearly Lord Rayleigh [1]. He introduced the concept of "resolving power" in application to spectral instruments and gave criteria of resolution. Rayleigh's resolution criterion, of course, refers to the distribution of two closely spaced monochromatic lines of equal intensity, and to the case when the instrument function is slowly determined by diffraction by a rectangular aperture, i.e., has the form

$$a(\lambda) = \frac{1}{S_0} \left(\frac{\sin \frac{\pi\lambda}{S_0}}{\frac{\pi\lambda}{S_0}} \right) \tag{III.1}$$

and is formulated as follows.

Two monochromatic lines of equal intensity are considered definitely resolved if the diffraction maximum of the first line coincides with the first diffraction minimum of the second. The intensity at the minimum of the total distribution in this case is about 81% of the maximum intensity.

It has been clear from the very outset that this definition is arbitrary and applicable only to the special case of two monochromatic lines of equal intensity, with the added assumption of a definite form imparted to the instrument function and the existence of a contrast sensitivity threshold (20% trough). The latter assumption is roughly correspondent with the properties of the eye, i.e., visual observations are being considered. With the development of spectroscopic techniques and broadening of the investigated region of the spectrum, instruments have begun to appear with instrument functions different from that considered by Rayleigh, and visual observation has gradually given way to more modern methods based on the use of photosensitive plates and such receivers as photocells, bolometers, counters, etc. A need has arisen for improving the concept of resolving power handed down by Rayleigh. The development of this concept has progressed mainly toward generalization in the sense of its application to instruments with an instrument function other than the one treated by Rayleigh (see, e. g., [72]). For this the resolution criterion was related to the depth of the trough; for example, it was assumed that the limit of resolution corresponded to a 20% trough [72-76] or 5% [77], in other words, all that was done was to adopt another value for the contrast sensitivity threshold. Sparrow [78] went beyond the others with this approach, regarding the resolution limit to be that position at which the trough between maxima "just vanished." This is expressed mathematically by the condition $d^2 f / d\lambda^2 = 0$, where $f(\lambda)$ is the observed contour.

Other resolution criteria were advanced, relating it, for example, to the ratio of the width of the instrument function to the separation between lines [79-81]. The influence of such factors as the intensity difference of the components [78-82], the effects of background [83-86], the finiteness of the line width [77, 82-87] have also been considered.

However, all of the proposed criteria are generalizations of the Rayleigh criterion, and only bring to bear on the discussion the different instrument functions. As far as accounting for the properties of the radiation receiver is concerned, essentially all of them retain the postulates of Rayleigh, since they have assumed the existence of a definite contrast sensitivity threshold. Nevertheless, for many receivers or sensors (an example will be given shortly) this assumption is not justified. We need, accordingly, in working out the criterion of resolution to take more detailed account of the factors (such as noise) that tend to limit resolution. Following this tack, it turns out that once again we are unable to consider the influence of the instrument function and measurement error independently, because, as a rule, the width of the instrument function is intimately connected with the magnitude of the random error. A suitable example here again is the purely slit-type instrument function.

As we know, if the slits are broadened the relative error diminishes rapidly. Consequently, for a given pair of lines the trough between lines will decrease as the slits are broadened, but at the same time the accuracy with which it measured will. Here again, then, we are faced with the problem of choosing optimum conditions of measurement, such that resolution will be optimal. It may be concluded from this that further development is required in the theory of resolving power, toward a more detailed accounting of the specific measurement errors causing resolution to be limited, i. e., a more detailed accounting of the properties of the receiver-sensors, the fluctuation processes in them that limit their capabilities. The problem before us is to seek optimum conditions of resolution.

On the other hand, the classical theory of resolving power is in need of yet another refinement. All classical criteria of resolution are connected in one way or another with observation of a dip or trough in the observed contour. As we are well aware, given some very finite separation between lines, this trough vanishes altogether. Consequently, the classical criteria lead to the notion of an absolute limit of resolution, which depends only on the instrument function (both the width and configuration of the line, if they are taken into account) and does not depend on the measurement error.

It appears that G. S. Gorelik [19] was the first to subject this aspect of resolution theory to critical analysis. He showed that, in spite of the disappearance of the trough, there are differences between the observed contour of one line and the observed contour of two lines (or any more complex structure), so that with sufficient accuracy in measuring the contours we have, in principle, the possibility of distinguishing these two cases. Developing this notion further, we can pose the question of reduction and its applicability. As we mentioned in the Introduction, in the absence of error in measuring the observed contour it is conceptually feasible to reproduce the true contour, hence to resolve any of its structure. Consequently, any method involving the observation of a trough is poor, because it does not utilize these conceptual possibilities; other methods must be sought for extracting information on the true contour, other criteria of resolution must be found.

The most general method can be built around the approach of L. A. Khalfin [25], which can be used, in particular, to determine the probability that a given structure is a doublet or triplet, etc. Examples of this approach are given in [3, 25]. It is very complex, however, and requires ponderous computations in the general case. It seems wise therefore to try and find simplified resolution criteria, which nevertheless will allow us to obtain in principle unlimited resolution with infinitesimal suppression of the measurement error.

36

As stated, the further development of the theory of resolving power demands refinement of the very concept of resolving power. The fact of the matter is that very often we interpret resolving power to mean some characteristic of the distorting properties of the instruments. A definite fixed spectral structure, for instance two monochromatic lines of equal intensity, is considered, and their resolution limit (as according to Rayleigh) is used to compare the distortion characteristics of various instruments. Naturally, such a characteristic of the distorting properties of the instrument is not complete, as they can only be completely described by the instrument function and accuracy characteristics of the instrument (of course for rough estimates, where the resolving power is usually used in this sense, it is still possible to use the width of the instrument function with a measure of success). On the other hand, such a determination is often unsuitable in practice in that it far from always possible to find the necessary doublet for the estimation of resolution.

In contrast with this, we will use the concept of resolution and resolving power in the more general sense as a concept characterizing the limiting capabilities of the instrument in the observation or measurement of a given structure of the spectrum. It is important to stress that this relates the terms resolution and resolving power not only with the attributes of the instrument itself and the conditions of measurement (its instrument function and measurement errors), but also with the properties of the object under investigation. Moreover, it is meaningful to speak of resolving power only when a definite means of extracting information is in mind. By this we mean, for example, observation of a trough in the observed spectrum or the application of reduction, processing by Khalfin's method, etc.

Summarizing, we can distinguish two main paths along which the theory of resolving power must develop. The first is associated with a more detailed and concrete accounting of the measurement error, hence a more detailed accounting of the properties of the radiation receiver. It is impossible, of course, to cover all variants of this problem. And we did not propose this as our aim. We will confine ourselves to the special case involving the resolving power of a typical infrared spectrometer. For this case we will take systematic account of the receiver noise, otherwise retaining the postulates of the classical resolution theory, i. e., relying on the observation of a trough. It will be seen in this example that only the measurement error is account for.

The second path of development involves searching for more up-to-date methods for extracting information from the observed contour. From these postulates we will consider the spectral approach to the problem of resolution and, with its help, try to find a simple criterion of resolution that does not lead to an absolute limit of resolution.

§1. Effect of Measurement Error on the Resolving Power of Infrared Spectrometers

The problem of the resolving power of spectral equipment is discussed in [88] for the case of absorption spectrum measurements; this work was subjected to a creditable critical analysis by I. V. Peisakhson [89]. Since, in our opinion, this critical analysis was not sufficiently comprehensive, we will consider it in more detail.

The first shortcoming of [88] (as, for that matter, of many other papers) is contained in the fact that it uses the Rayleigh criterion, which is suitable only for a certain special case. The authors use the Rayleigh criterion in the formulation of $I_{min}/I_{max} = 0.8$, without regard for the absolute values of I_{min} and I_{max}. At large absorption, i. e., small absolute values of these intensities, increased resolution is obtained relative to the emission spectra.

The critique of [88] in [89] briefly consists in the following: It is indicated that at large absorption the quantities I_{min} and I_{max} themselves will be small, while the relative error of their measurement will be large. As a result, instead of enhanced resolution it will in general be difficult to detect a trough. On this basis the opinion is expressed that for absorption spectra the resolution criterion must be based on inspection of the absorption rather than the transmission spectrum.

However, the main failing of [88] lies not with the use of the transmission spectrum but with the fact that the Rayleigh criterion is used, which presupposes, as mentioned, visual observation. The applicability of this criterion must be accepted or rejected on the basis of an analysis of the properties of the measurement errors inherent in a given piece of equipment. This criterion is inapplicable to the majority of modern instruments.

The shortcomings of [88] are not confined to this, however. A second deficiency is that the features of the absorption spectra, which are essential to the theory of resolving power, are not taken into account. In fact, the authors consider the case when the observed width of the bands is determined by the instrument function, and here again they analyze the case of large observed absorption values. However, the magnitude of the absorption cannot be greater than unity, so that the absorption bands cannot possibly be treated as delta functions. This means that if we consider the bands to be considerably narrower than the instrument function by virtue of retaining the total absorption, the observed absorption of such bands will be very small, in fact smaller the broader the instrument function. The observed absorption can be increased in this case only by increasing the total absorption, for example by increasing the thickness of the sample. But then the band will broaden, since the peak absorption rapidly approaches unity, and beyond this only its wings can be increased. Hence we obtain a large observed absorption only when the width of the band is comparable with the width of the instrument function. This means that for absorption spectra the theory of resolving power must be formulated for bands of finite width.

It follows from this that the results of [88], including the assertion that in absorption spectra the resolving power of a given instrument at large absorption can be greater than in emission spectra, must be considered erroneous.

In [89] a systematic analysis of the effect of measurement error is also given, but then it is only pointed out that they need to be taken into account. Furthermore, allowance is not made for the abovementioned characteristic of absorption spectra. Consequently, an analysis of the resolving power of instruments for the measurement of absorption is essentially lacking. In particular, the problem of the actual resolving power of infrared spectrometers remains open to inquiry. The well-known criteria of resoltuion are inapplicable to these instruments, since their instrument function is determined principally by the slits, whereas the usual criteria of resolution are intended for instrument functions of another form, and besides, what is particularly important, in the given case it is impossible to proceed without regard for the random error of measurement (noise), for in these instruments it is a very decisive factor. In this connection we will consider the resolving power of infrared spectrometers, orienting toward the standard types of instruments. We will solve a problem which in a certain sense is analogous to the Rayleigh problem of discrimination between two closely spaced identical[*] lines, but for absorption bands of finite width and with consideration for the specific properties of the optical receiver and instrument function, as well as their interrelation. The results of this analysis will aid in equal measure in evaluating the capabilities of modern infrared instrumentation in the sense of distinguishing complex structures in the absorption spectrum, just as the Rayleigh resolving power aids in the evaluation of the capabilities of the eye in distinguishing various objects.

Since clearly the majority of bands encountered in practice have a configuration intermediate between dispersion and Gaussian types, these two structures may be regarded as limiting cases and used to evaluate the resolution for the entire range of band configurations. Consequently, for our test structures we will use the following: two identical bands of Gaussian configuration; two identical bands of dispersion configuration. The absorption coefficient of the sample in the first case is specified in the form

$$\varepsilon_1(\lambda) = e^{-\frac{4\lambda^2 \ln 2}{\gamma^2}} + e^{-\frac{4(\lambda - \Delta)^2 \ln 2}{\gamma^2}}, \tag{III.2}$$

and in the second case as

$$\varepsilon_2(\lambda) = \frac{1}{1 + \frac{4\lambda^2}{\gamma^2}} + \frac{1}{1 + \frac{4(\lambda - \Delta)^2}{\gamma^2}}, \tag{III.3}$$

where γ is the width of the individual band at the half-power points. Here the shape of the true absorption curve is more complex and is given by the expression

---

[*]In other words, bands having the same configuration, width, and peak intensity.

$$x(\lambda) = 1 - e^{-\varepsilon(\lambda)d}, \tag{III.4}$$

where d is the thickness of the absorbent layer.

We will calculate the resolution limit for a typical infrared instrument. The required characteristics of such an instrument as applied to our problem are discussed in [31]. We will limit ourselves, however, to just the simplest case, assuming that the instrument function is triangular in form with a half-width equal to S, i. e., the width of the monochromator slits (slits of equal width).

To characterize the random error of measurement we use the equation

$$\overline{\delta x^2} = E/\tau S^4.$$

Criterion of Resolution. In this section we will be concerned only with the influence of the measurement error, so that, just as in the classical theory of resolving power, the resolution criterion will be based on discrimination of a trough between peaks in the direct record of the total absorption curve $x(\lambda)$, i. e., we will not assume reductions to an ideal instrument or any other sort of manipulations.

In order to explicitly account for the influence of the measurement error it is natural to adopt as the measure of resolution, just as in electronic engineering, the ratio of the depth of the trough to the mean square error:

$$\rho = \frac{\Delta x}{\sqrt{\overline{\delta x^2}}}, \tag{III.5}$$

where $\Delta x = x_{max} - x_{min}$.

We will consider a structure resolved when for that structure ρ is larger than some threshold value ρ_{lim}. The choice of ρ_{lim}, of course, is arbitrary and depends on what degree of reliability we wish to impart to the trough between peaks. We may, for example, consider a structure resolved for which the depth of the absorption trough is greater than or equal to the mean square error in measuring the absorption, i. e., $\rho \gg 1$. It is necessary for determining the resolving power R to find at what distance Δ_{lim} between bands $\rho = \rho_{lim}$. Then the resolving power is determined from the relation

$$R = \lambda/\Delta\lambda = \lambda/\Delta_{lim}. \tag{III.6}$$

Optimum Conditions of Resolution. With a fixed distance between bands the value of ρ in our case will depend on two more parameters, which the experimenter can vary throughout his work: the thickness d of the absorbent layer and the slit width S (the scan rate will be assumed fixed, while the time constant of the instrument is chosen so as to eliminate additional distortion due to scanning). To determine the maximum resolving power attainable with a given instrument we need to find the maximum $\rho(d, S)$ with respect to these parameters. Such a maximum does exist, since broadening of the slits, on the one hand, produces an increase in the observed width of the band, hence decreases Δx; on the other hand, according to an equation given in the Conclusion, it tends to decrease $\overline{\delta x^2}$. In the extreme cases of very large and very small slits the resolution of the bands will infinitely worse. It is readily apparent that ρ also has a maximum with respect to d.

Choice of Layer Thickness. To find the optimum conditions we need at the same time to find the maximum of $\rho(d, S)$ with respect to the two parameters d, S. However, as calculations show, we can, without incurring too much error, find the maximum with respect to these two parameters independently. Since the random error does not depend on the magnitude of the absorption we can look for an optimum layer thickness d from the condition of a maximum trough Δx in the true absorption curve.

Figure 11 shows the depth of the trough Δx in the true absorption curve, obtained by direct calculation from Eq. (III.2), as a function of the peak absorption x_0 of the overall true absorption curve for several values of the parameter Δ/γ in the case of two Gaussian bands. Figure 12 gives the same dependence for two dispersion bands.

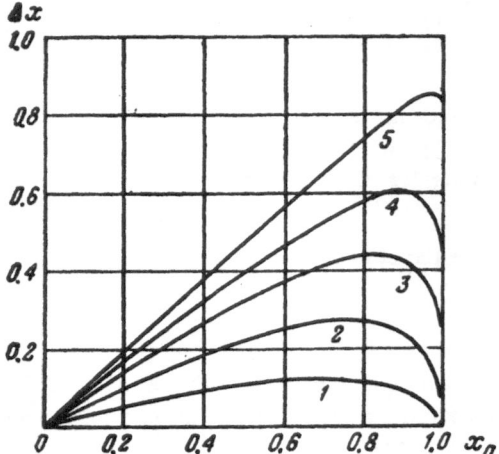

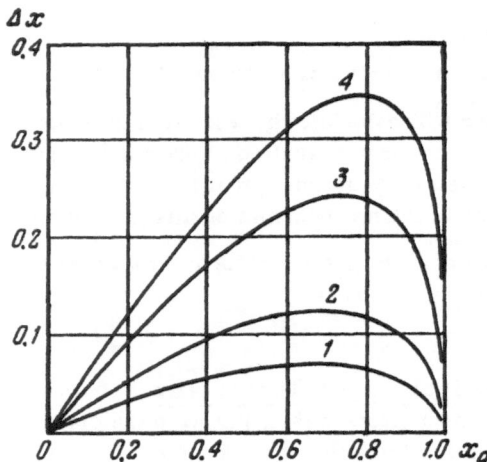

Fig. 11. Dependence of the trough depth Δx in the true absorption curve on the peak absorption x_0 of this curve for two Gaussian bands. Δ/γ = 1) 1.2; 2) 1.44; 3) 1.68; 4) 1.92; 5) 2.4.

Fig. 12. Dependence of the trough depth Δx in the true absorption curve on the peak absorption x_0 of this curve for two dispersion bands. Δ/γ = 1) 1.0; 2) 1.2; 3) 1.6; 4) 2.0.

Allowance for the influence of the slit width yields values of d_m and x_{0m} somewhat larger than those corresponding to the peaks of the curves in Figs. 11 and 12. However, this only causes a slight increase in ρ (5% for our range of values of Δ/γ). We will assume in our future calculations, therefore, that the optimum values of x_{0m} and the layer thickness d_m correspond to the peaks of the curves in Figs. 11 and 12.

Figure 13 shows the dependence of the values thus obtained for x_{0m} and d_m and their corresponding values of Δx on the value of Δ/γ for the case of Gaussian and dispersion curve configurations.

The optimum values of x_{0m} turn out to be rather large, increasing considerably as Δ/γ is increased; for $\Delta/\gamma > 1.5$ they go outside the normally recommended working absorption interval, i.e., 20-70% absorption. This applies, of course, to the true absorption values; the observed values will be somewhat less under optimum conditions.

<u>Choice of Slit Width.</u> The effect of a slit-type instrument function on the absorption curve has been calculated by graphical integration, so that the accuracy of the calculations is a few percent, which is nevertheless satisfactory for our purposes.

Figure 14 gives the dependence on Δ/γ of the optimum slit width, expressed in units of $\gamma/2$, i.e., the ratios $2S_m/\gamma$ for the two cases indicated above (Gaussian and dispersion band configurations). Also shown are the values of $\Delta x \cdot 4S^2/\gamma^2$ corresponding to S_m and defining ρ, since

$$\rho = \frac{S^2}{\gamma^2}\,\Delta x U_1^{-1/2}, \tag{III.7}$$

where

$$U_1 \doteq E/\tau\gamma^4.$$

It is easy to see from Fig. 14 that for a given value of Δ/γ the values of $2S_m/\gamma$ are very near Δ/γ the values of $2S_m/\gamma$ are very near Δ/γ for Gaussian and dispersion bands.

In rough terms, S_m can be chosen simply from the conditions $2S_m/\gamma = \Delta/\gamma$, or $S_m = \Delta/2$. This means that with $\Delta/\gamma \simeq 1.1$ such a choice of S_m will make the value of ρ 20% less than optimum for the two Gaussian

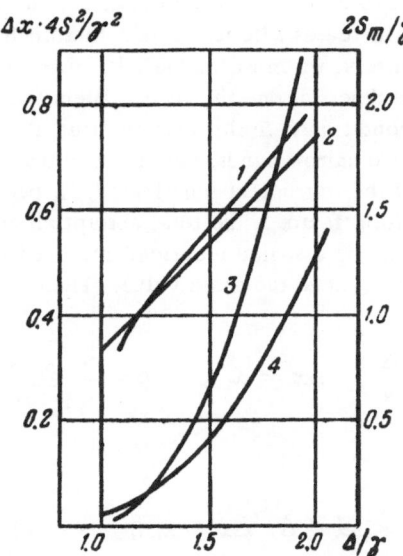

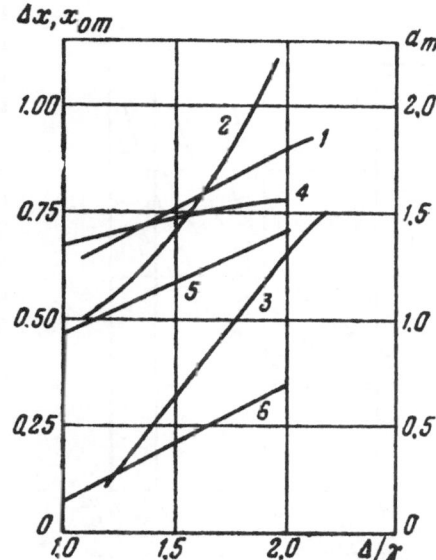

Fig. 13. Dependence of the optimum x_{0m} and d_m (arbitrary units of measure) and the corresponding Δx on the ratio Δ/γ. Two Gaussian bands: 1) x_{0m}; 2) d_m; 3) Δx; two dispersion bands: 4) x_{0m}; 5) d_m; 6) Δx.

Fig. 14. Dependence of the optimum values of the slit width, relative to $\gamma/2$, and the corresponding values of $\Delta x \cdot 4S^2/\gamma^2$ on Δ/γ. 1,2) $2S_m/\gamma$ for Gaussian and dispersion bands, respectively; 3,4) $\Delta x \cdot 4S^2/\gamma^2$ for the same cases.

bands, while for the two dispersion bands with $\Delta/\gamma \approx 1.0$ the value of ρ will be 17% less than optimum. Beginning with $\Delta/\gamma = 1.25$, over our entire interval of values of Δ/γ the decrease in ρ will never the more than 5 to 7% with this choice of S_m.

Resolution Limit. We agreed arbitrarily on the limit of resolutions as that distance between bands at which the variable ρ has a definite assigned value ρ_{lim}. Proceeding from this criterion of resolution and making use of curves 3 and 4 in Fig. 14, along with Eq. (III.7), we can now calculate the dependence of Δ_{lim}/γ on the parameter U_1. The resolving power R of the instrument is related to Δ_{lim}, which depends only on U_1, by the expression

$$R = \frac{\lambda}{\Delta_{lim}(U_1)} .$$

Since we have calculated the optimum conditions of measurement, the resolving power thus obtained is maximal for the given instrument.

Figure 15 shows how the threshold quantity Δ_{lim}/γ depends on the parameter U_1. It is clear from this that the ultimately resolvable distance Δ_{lim} for U_1 does not depend strongly on U_1. The ultimately resolvable distance between bands as determined above will tend to a constant as the error tends to zero. This is reasonable, for if bands of finite width are spaced closely enough together the trough between peaks will eventually vanish. Consequently, for bands of a given width γ it is to little avail to make E larger than a certain value when using such a criterion of resolution, because the resolution will change very little. On the other hand, for instruments of inferior quality (large E) and for structures consisting of narrow bands (small γ), i.e., for large U_1, the function $\Delta_{lim}(U_1)$ will become significant; in this case the resolution will be improved by decreasing E.

Resolution of Two Narrow Bands. It is instructive to consider the limiting case of resolution of two narrow bands, i.e., those for which $\gamma \ll \Delta$, or $\Delta/\gamma \gg 1$. In this case the optimum value of the peak

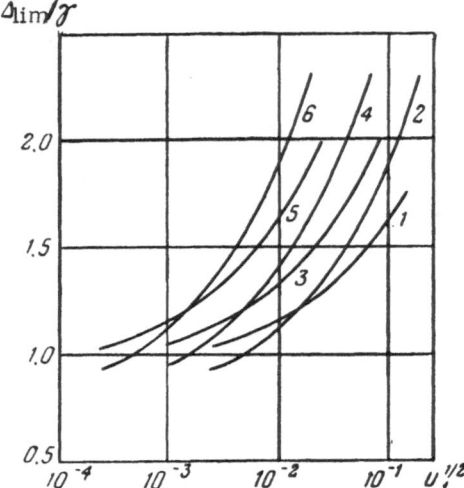

Fig. 15. Ultimately resolvable distance $\Delta_{\lim}$ between bands, relative to the width of the band γ, as a function of the parameter $U_1^{1/2}$ for different values of $\rho_{\lim}$. 1, 3, 5) $\rho_{\lim} = 1, 3, 10$, respectively for two Gaussian bands; 2, 4, 6) the same, for two dispersion bands.

absorption tends to unity as Δ/γ increases, and the optimum layer thickness increases without bound. Inasmuch as we cannot increase d infinitely, we must for the sake of practicality consider a fixed thickness d, one that is consistent with definite experimental conditions. In this case the true absorption curve will represent two narrow bands, the observed one will be a superposition of the instrument functions, i.e., two triangles. The only important factor is the total absorption of the band, and the bands can be assumed identical, provided their integrals $\int x(\lambda)\,d\lambda = x_\infty$ have the same value. Then

$$\text{for } 0 < S < \frac{\Delta}{2}, \quad \Delta x = x_{0.\text{obs}}; \quad \rho = \frac{x_{0.\text{obs}}\,S}{\sqrt{E/\tau}} = \frac{x_\infty S}{\sqrt{E/\tau}};$$

(III.8)

$$\text{for } \frac{\Delta}{2} < S < \Delta, \quad \Delta x = x_{0.\text{obs}} - 2x\left(\frac{\Delta}{2}\right)$$

$$= x_{0.\text{obs}}\left(\frac{\Delta}{S} - 1\right), \quad \rho = \frac{x_\infty}{\sqrt{E/\tau}}(\Delta - S).$$

It is readily apparent that

$$S_m = \Delta/2. \tag{III.9}$$

It is important to realize that this equality is identical to the condition that we obtained as approximate for a limited interval of Δ/γ. It may be considered on this basis that the condition $S_m = \Delta/2$ is applicable to the entire interval of $\Delta/\gamma > 1.25$.

The value obtained for S_m results in the following resolution limit:

$$\Delta_{\lim} = \frac{2\rho_{\lim}}{x_\infty}\sqrt{E/\tau}, \tag{III.10}$$

whence the resolving power under optimum conditions is

$$R = \frac{\lambda}{\Delta\rho} = \frac{\lambda x_\infty}{2\lambda_{\lim}\sqrt{E/\tau}}, \tag{III.11}$$

i.e., it is directly proportional to the total absorption of the band and inversely proportional to the assumed $\rho_{\lim}$ and the mean square random error $(E/\tau)^{1/2}$ (E is a parameter of the instrument, characterizing its random error).

§2. Spectral Approach to the Resolution Problem

In this section we will attempt to exercise a departure from the conventional method of structure discrimination in a spectrum, which is related to the observation of a trough between peaks, and find alternate methods of resolution. It has been shown that with the use of the classical criterion of resolution the natural requirement of an infinite resolving power in the absence of error is not fulfilled. Our task therefore will be to find methods of observation that satisfy this requirement. The only means of deriving such a method of observation (other than the method of L. A. Khalfin indicated at the beginning of the chapter) consists in

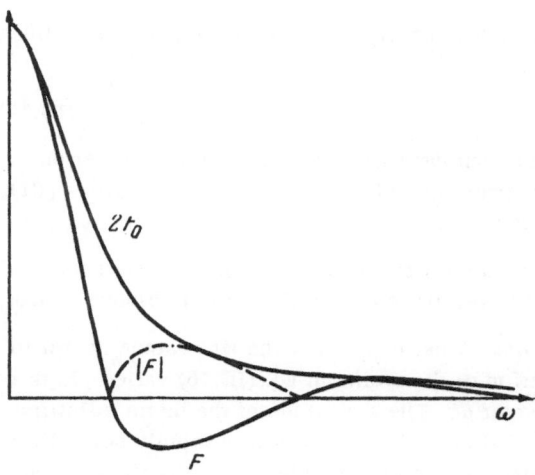

Fig. 16. Form of the Fourier transform of the observed contour of two identical lines.

applying reduction. In actuality, using reduction we approach the true distribution and, hence, can resolve its structure to the extent permitted by the increased noise of reduction.

Here also we can pose the question of what are the optimum conditions of measurement and reduction, then determine the limiting capabilities of the instrument with such a method of observation. It is clear that the resolution limit in this case will be restricted by the measurement error. However, both the method of Khalfin and reduction are fairly complex in actual practice, as well as in theoretical investigation. We will begin with a simple case, in which we can pick out the fundamental details of the problem. We note for this purpose that reduction is particularly simple to treat by means of the "spectral approach" (see the Introduction). In fact, reduction leads to division of $F(\omega)$ by $A(\omega)$ or by its approximate value. It may be considered that the presence of structure in $\varphi(\lambda)$ will lead to definite characteristics in the form of $\Phi(\omega)$. These characteristics ought to be manifested also in $F(\omega)$ as affected by $A(\omega)$. Since this influence means multiplication by $A(\omega)$, it is hoped that we can draw some sort of conclusion as to $\varphi(\lambda)$ on the basis of $F(\omega)$.

Consider $\varphi(\lambda)$, which consists of bands with identical form:

$$\varphi(\lambda) = \sum_j c_j \varphi_0 (\lambda - \Delta_j). \tag{III.12}$$

A displacement by the amount Δ_j, as we have seen, is equivalent to multiplication by $e^{i\omega \Delta_j}$ in the Fourier transform. Consequently,

$$\Phi(\omega) = \sum_j c_j \Phi_0(\omega) e^{i\omega \Delta_j} = \Phi_0(\omega) \sum_j c_j e^{i\omega \Delta_j}, \tag{III.13}$$

i. e., the Fourier transform of the total distribution consisting of identically shaped bands is equal to the Fourier transform of one component multiplied by the trigonometric sum $\sum_j c_j e^{i\omega \Delta_j}$. The form of this sum depends on the number and relative intensity of the components of $\varphi(\lambda)$. For simple cases it has a very simple and graphic form. Thus, for example, it is readily shown that for two bands of identical intensity

$$\varphi(\lambda) = \varphi_0(\lambda + \Delta/2) + \varphi_0(\lambda - \Delta/2), \quad \Phi(\omega) = \Phi_0(\omega) \cos \omega \Delta/2. \tag{III.14}$$

The effect of the instrument function amounts to multiplying $\Phi(\omega)$ by $A(\omega)$:

$$F(\omega) = F_0(\omega) \sum_j c_j e^{i\omega \Delta_j} = \Phi_0(\omega) A(\omega) \sum_j c_j e^{i\omega \Delta_j}. \tag{III.15}$$

For two identical bands

$$F(\omega) = F_0(\omega) 2 \cos \omega \Delta/2 = 2\Phi_0(\omega) A(\omega) \cos \omega \Delta/2. \tag{III.16}$$

The graph of $F(\omega)$ for this case is shown in Fig. 16. We see that the graph of $F(\omega)$ is a cosinusoid, inscribed in $2F_0(\omega)$; hence $|F(\omega)|$ has a series of characteristic troughs that dip to zero, repeating in equal intervals. From the frequency ω_1 corresponding, for example, to the first zero, we readily find the spacing between lines:

$$\Delta = \pi/\omega_1. \tag{III.17}$$

These typical features of the Fourier transform of complex contours were used long ago by Michelson in his investigations of the structure of lines by means of a twin interferometer [90], which, as we are aware [91], at once give the Fourier transform of the distribution being investigated.

It is important for us to show that a resolution criterion based on observation of a trough in the Fourier transform of the distribution under scrutiny will lead to an infinite resolving power in the total absence of error.

Let us consider for simplicity the case of two identical lines. What happens as the lines come nearer to one another? The factor $F_0(\omega)$ remains unchanged, only the period of the cosine in Eq. (III.16) varies. In this case the first trough and all subsequent ones will shift toward larger ω. The amplitude of the hump following the first trough is used to estimate the presence of structure in the line and diminishes very rapidly here. However, the trough at no time decays, so that in the absence of error we can in principle detect the troughs and, consequently, the fact of structure present in the line, and we can even measure the separation between lines for any spacing, no matter how near. In practice, of course, as the lines merge together the difficulty of their resolution accumulates very rapidly; this is because with a certain reduction of the hump and the presence of error we are no longer able to detect it.

A special explanation is required for the cases of diffraction and slit instrument functions. The Fourier transform of the diffraction function $a(\lambda)$ has the form of an isosceles triangle:

$$A(\omega) = \begin{cases} 1 - \dfrac{S_0}{2\pi}|\omega| & |\omega| \leqslant \dfrac{2\pi}{S_0}, \\ 0 & |\omega| > \dfrac{2\pi}{S_0}. \end{cases} \tag{III.18}$$

It can be seen that $A(\omega)$ goes to zero beyond the limits of some interval of ω. Rautian [21] showed, however, that such an expression corresponds to solution of the diffraction problem by the approximate method of Kirchhoff. Rigorous solution produces a smooth decay in $A(\omega)$. Consequently, the conclusion of an infinite resolving power is also applicable in principle to this case, although in practice we must clearly be aware that the decrease of $A(\omega)$ is very abrupt here, so that it becomes extremely difficult to penetrate beyond the barrier of the diffraction limit.

It is interesting to find out to what the Rayleigh resolution criterion corresponds. Rayleigh investigated monochromatic lines, so that

$$\Phi_0(\omega) = \text{const},$$
$$F(\omega) = \text{const} \cdot 2\left(1 - \frac{S_0}{2\pi}|\omega|\right)\cos\frac{\omega\Delta}{2}. \tag{III.19}$$

The limiting resolvable distance according to Rayleigh is $\Delta = S_0$. Here the first zero corresponds to the frequency $\omega_1 = \pi/S_0$ for which $A(\omega) = 1/2$. Figure 17 shows $|F(\omega)|$ for a line separation Δ_R corresponding to the Sparrow resolution limit [78] (curve S). The straight line gives the shape of $A(\omega)$. As we see, neither of these criteria fully utilize the possibility of resolution. In the

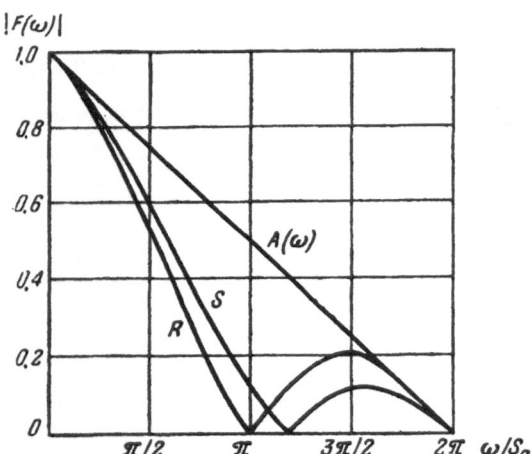

Fig. 17. Absolute value of the Fourier transform of the observed contour of two identical lines for the case of the Rayleigh resolution limit (curve R) and the Sparrow resolution limit (curve S). $A(\omega)$ is the Fourier transform of the instrument function.

limit as the accuracy is increased accordingly, we could, on the basis of the first trough in $|F(\omega)|$ with a purely diffraction type $A(\omega)$ as described by (III.18), resolve lines half the distance apart distinguished by the Rayleigh criterion.

The other instrument function that requires special consideration is the purely slit type instrument function. For this case $A(\omega)$ itself has zeros, which can concur with the zeros of $\Phi(\omega)$. However, this danger can be avoided, since it is always possible to distinguish the zeros of $A(\omega)$ from the zeros (or troughs) of $\Phi(\omega)$. This can be done by changing the slit width; then the zeros of $A(\omega)$ will be shifted by a known amount, while the zeros of $\Phi(\omega)$ will remain in the same places. Moreover, knowing the slit width, calculation of the position of the zeros of $A(\omega)$ is quite straightforward.

Already the analysis of these uncomplicated examples reveals that investigation of the Fourier transforms of the observed distributions can be very useful in resolving the structure of spectra. This was realized even by Michelson [90]. In recent times the reawakening and development of Michelson's method are causing such techniques again to find application. Connes [92] describes the application of zeros in the interference pattern obtained with a twin-wave interferometer for the resolution of doublets; in [93] expressions are derived for $\sum_l c_l e^{i\omega\Delta_l}$ for a number of simple cases; in [94] it is proposed that these methods be used for the analysis of rotational spectra. Their application, however, is not limited just to twin-wave interferometers, which yield $F(\omega)$ directly; it is also possible to take the Fourier transform of the recorded $f(\lambda)$ obtained on an ordinary spectral device and to use the techniques described here with success. The limiting capabilities of instruments will be determined by their accuracy characteristics, and from this point of view they are amenable to comparison.

Let us see how the presence of measurement error limits the possibilities of resolving spectral structure in terms of a trough in the Fourier transform. We will examine a simple case. Let us measure a doublet consisting of two identical lines. In order to circumvent complicated expressions, we will assume that the shape of the observed bands is the dispersion type, with width γ. We will not be concerned with the problem of what is determined by this width—the width of the instrument function or of the line itself. In this case the total observed distribution becomes

$$\varphi(\lambda) = \varphi(0) \frac{1}{1 + \dfrac{4(\lambda - \Delta/2)^2}{\gamma^2}} + \varphi(0) \frac{1}{1 + \dfrac{4(\lambda + \Delta/2)^2}{\gamma^2}} + \xi(\lambda). \tag{III.20}$$

Here Δ is the distance between lines, $\xi(\lambda)$ is the measurement error.

For the Fourier transform we obtain

$$F(\omega) = \frac{\pi}{2}\varphi(0)\gamma e^{-\frac{\gamma}{2}|\omega|} \cos\frac{\omega\Delta}{2} + X(\omega), \tag{III.21}$$

where $X(\omega)$ is the Fourier transform of $\xi(\lambda)$.

We now find the amplitude of the first hump (disregarding error):

$$\frac{dF}{d\omega} = \pi\varphi(0)\gamma\left(-\frac{\gamma}{2}\right)e^{-\frac{\gamma}{2}|\omega_m|}\cos\frac{\omega_m\Delta}{2} - \pi\varphi(0)\gamma\frac{\Delta}{2}e^{-\frac{\gamma}{2}|\omega|}\sin\frac{\omega_m\Delta}{2} = 0.$$

Hence

$$\omega_m = -\frac{2}{\Delta}\arctan\frac{\gamma}{\Delta}, \quad F(\omega_m) = \pi\varphi(0)\gamma e^{-\frac{\gamma}{\Delta}\left|\arctan\frac{\gamma}{\Delta}\right|}\frac{1}{\sqrt{1 + \gamma^2/\Delta^2}}. \tag{III.22}$$

As was done in §1, we will consider the resolution limit to be that case when the error comprises a definite part of the height of the first hump. Then the resolution condition becomes

$$\frac{F(\omega_m)}{\sqrt{\overline{X^2(\omega_m)}}} = \rho_{lim} \tag{III.23}$$

The choice of ρ_{lim} here again depends on what degree of confidence in the resolution is desired.

Under real conditions, naturally, we will always record the spectrum over a finite interval of Λ. Then instead of (III.21) we must obtain from the theorem of the product spectrum the convolution of $F(\omega)$ and the function $(\sin \Lambda\omega/2)/(\Lambda\omega/2)$. This brings about a certain diminution of the humps and a displacement of the zero positions.

For $\overline{X^2}(\omega)$ we can obtain [27]

$$\overline{X^2}(\omega) = \pi\Lambda\Psi(\omega). \tag{III.24}$$

Here $\psi(\omega)$ is the spectral power of the error (Fourier transform of the correlation function). It is assumed here that $\Lambda \gg \tau\nu$, where τ is the error correlation time. The situation here is approximately the same as in the choice of a time constant for recording of the spectrum, since the systematic distortions also govern the convolution, while the mean square random distortion is proportional to Λ. By analogy with an earlier result we should consider it suitable to use comparatively large Λ, such that the systematic distortions will be small. Neglecting them, we find

$$\frac{\pi\varphi(0)\,\gamma\exp\left[-\dfrac{\gamma}{\Delta}\left|\arctan\dfrac{\gamma}{\Delta}\right|\right]}{\sqrt{1+\gamma^2/\Delta^2}} = \rho_{lim}\sqrt{\pi\Lambda\Psi(\omega_m)}. \tag{III.25}$$

Usually the error spectrum turns out to be uniform. Assuming therefore that $\psi(\omega) = \psi_0 = $ const., we obtain

$$f\left(\frac{\gamma}{\Delta_{lim}}\right) = \frac{\exp\left[-\dfrac{\gamma}{\Delta_{lim}}\left|\arctan\dfrac{\gamma}{\Delta_{lim}}\right|\right]}{\sqrt{1+\gamma^2/\Delta_{lim}^2}} = \rho_{lim}\sqrt{\frac{\Lambda\Psi_0}{\pi\varphi^2(0)\,\gamma^2}}. \tag{III.26}$$

This equation contains the dependence of Δ_{lim}/γ on the parameters of the instrument. Noting that

$$\overline{\xi^2} = \frac{1}{2\pi}\int_{-\infty}^{\infty}\Psi(\omega)\,d\omega = \frac{\Psi_0\Omega_{eff}}{\pi},$$

where Ω_{eff} is the effective width of the error band, we see that the ultimately resolvable separation Δ_{lim} is determined by the quantity

$$\frac{\Lambda\Psi_0}{\pi\varphi^2(0)\,\gamma^2} = \frac{\Lambda\overline{\xi^2}}{\varphi^2(0)\,\gamma^2\Omega_{eff}},$$

i.e., essentially by the relative accuracy in measuring $\varphi(\lambda)$.

Figure 18 shows the function $f(\gamma/\Delta_{lim})$ on the left-hand side of (III.26). The limiting resolvable separation is determined by the intersection of this curve with the horizontal line at the height $\rho_{lim}[\Lambda\psi_0/\pi\varphi^2(0)\gamma^2]^{1/2}$. We see that as the relative measurement error tends to zero Δ_{lim} also approaches zero.

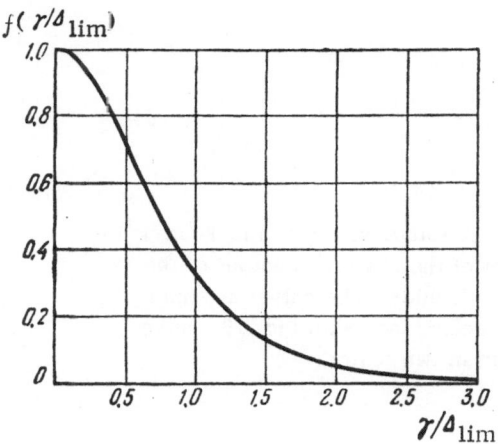

$f(\gamma/\Delta_{lim})$

Fig. 18. Curve for determining the resolution limit with observation of the trough in $F(\omega)$ for the case of two identical dispersion bands.

In the case of lines with a different shape we obtain qualitatively the same picture, except that the form of the function $f(\gamma/\Delta_{lim})$ will depend on the shape of the lines. Accordingly, the steeper the descent of $f(\omega)$, naturally the greater will be the increase in accuracy required for resolution of the same interval Δ_{lim}. For the shapes of $f(\lambda)$ encountered in practice $F(\omega)$ normally descends very steeply, so that increased resolution can only be obtained at the cost of a considerable increase in accuracy.

It is interesting to find out whether the accuracy attainable in ordinary measurements is such as to permit increased resolution, using the method described above. The following experiment was devised to ascertain this.

On a twin-wave infrared spectrometer, type AIKS-F-4 [95],the following doublet in the spectrum of gaseous ammonia was recorded: the bands ν_1 = 892.10 cm^{-1}, ν_2 = 887.96 cm^{-1}. The bands comprising the doublet are very narrow, so that the width of the bands and resolution of the doublet were determined mainly by the width of the instrument function: The latter varied with variation of the slit width S, the trough between peaks in the record gradually vanishing. These records are reproduced in Fig. 19.

In parallel with the recording on paper the spectrum was also recorded on punched tape (see Chapter IV), after which the Ural electronic computer was used to compute the Fourier transform of the recorded contours. The curves in Fig. 20 show the function $|F(\omega)|$ corresponding to the records of Fig. 19. It can be shown from these curves that the trough in $F(\omega)$ is definitely detectable in those cases when no trough at all is detectable in the original records. The fact that the trough does not dip to zero is associated with the fact that the components of the doublet are by no means identical, and perhaps with noise effects. The cited curves can even be used to measure a doublet separation which turns out to be approximately 4.0 cm^{-1} when the true value is 4.14 cm^{-1}.

We see, then, that under ordinary measurement conditions the use of Fourier transforms yields a considerable enhancement of resolution and provides a limit that is associated with vanishing of the trough in the observed contour. With a commensurate improvement in the measurement techniques, clearly, even better resolution will be possible.

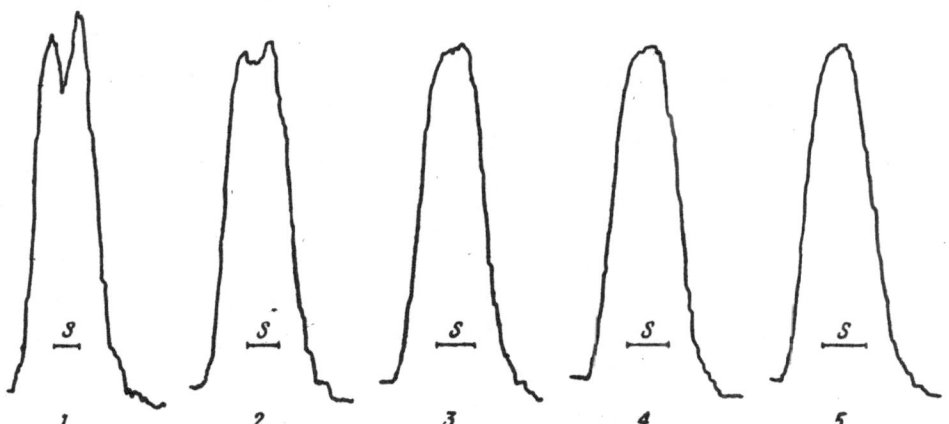

Fig. 19. Records of ammonia doublet ν_1 = 892.10 cm^{-1} and ν_2 = 887.96 cm^{-1} on the AIKS-F-4 instrument with NaCl prism. 1) S $\simeq$ 0.220 mm; 2) S $\simeq$ 0.270 mm; 3) S $\simeq$ 0.320 mm; 4) S $\simeq$ 0.340 mm; 5) S $\simeq$ 0.360 mm.

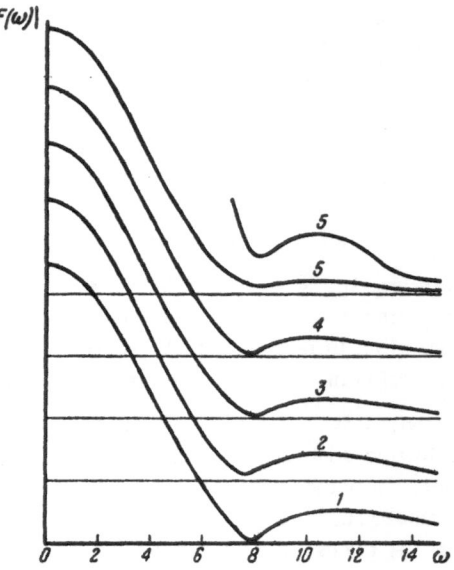

Fig. 20. Absolute value of the Fourier transform of the observed contour of the ammonia doublet. The curves are numbered in accordance with Fig. 19; curve 5 is given in two scales.

The example given here is only a special case of course, but it gives us an idea of the opportunities afforded by the Fourier transform for analysis of complex spectra with conventional instruments. Above all, we note that a resolution criterion based on the observation of a trough in the Fourier transform of the measured spectrum makes it possible to pose with validity the problem of a resolution limit and the effects of measurement error on this limit. In certain instances this promises a substantial increase in resolution owing to enhancement of measurement precision, especially in those cases when the structure of the investigated spectrum is known a priori, as for example in the case when one is measuring Zeeman structure or isotopic splitting of lines, etc. Of course, this method is particularly beneficial for the analysis of spectra consisting of bands identical in shape, but even for other cases application of the Fourier transform facilitates resolution of spectral structure, although the applications may require more complex methods of processing.

CHAPTER IV

APPLICATION OF DIGITAL COMPUTERS
FOR THE AUTOMATIC PROCESSING
OF INSTRUMENT DISTORTION [31, 71]

§1. Application of Computers for Automation of Reduction

In Chapter II we examined the methods of reduction to an ideal instrument and saw that the hypothetical solution of this problem is a familar one. However, its practical solution falters over a whole series of purely computational difficulties. The need for carrying out painstaking and laborious calculations at present seriously curtails the application of reduction. For this reason only approximate methods, such as the method of Rayleigh [2], generalized by Bracewell [14], or methods that assume a known simple form of the measured distribution and instrument function, applicable only in rare situations [48, 51, 69, 96], have any sort of reputation for more or less widespread application. These methods, however, also require rather tedious processing of the spectra. Consequently, we still are far removed from the practical application of reduction in anything like a general case. Nevertheless, the feasibility of a practical solution to this problem exists right now and involves implementation by modern computer techniques.

In our institute we have had the opportunity to work with a general-purpose digital computer, the "Ural" [97]. This is a single-address machine with fixed-point arithmetic, capable of executing 100 operations per second. The main memory has a working capacity of 1024 digits. The machine uses a punched tape input.

§2. Numerical Code System for Recording of Spectra

The greatest difficulties in using digital computers for the processing of spectra arise in connection with getting the spectral data into the machine. The problem is that all spectral instruments provide the recorded spectrum in continuously varying quantities, for example as the output voltage of an amplifier or in the linear displacement of an optical wedge. The spectrum in this case is usually recorded by a stylus on paper tape or by a "flying spot" on photographic paper. However, in computer equipment the spectrum must be fed in as a series of discrete numbers recorded in coded form on punched tape, punched cards, or magnetic tape. This being the case, the process of inputting data into the machine is a very difficult and tedious operation and does not ensure sufficient accuracy. It is necessary to measure manually the spectral distribution recorded, say, on graph paper tape, where the number of points at which the intensity is measured is sometimes as much as several hundred; all of these data need to be recorded on special forms, then translated manually to punched tape or punched cards. It goes without saying that these operations consume a good deal of time, and it would indeed be futile to hope for the avoidance of processing errors. A simple calculation shows that with manual data processing, the preparation for machine input of a spectrum measured at 200 points requires approximately two or three times the total operating time. Such time expenditures for the input of data, of course, largely rule out the advantages of the digital computer, and in the case of relatively simple processing of the spectrum it renders the use of such equipment inadvisable. It is clear, then, that the rational utilization of digital computers requires the creation of a system that would automate recording of the spectra in an appropriate code directly in the instrument recording process.

Such a system was designed and built according to our specifications by the computer service group of our institute, under the direction of Chief Engineer B. E. Marchuk.*

The system designed for recording of the spectra in machine code has the following basic components[98]:

1. A digital voltmeter, which converts the continuously varying signal delivered to its input into digital code. The measured voltage is represented in binary code by a nine-place number. This corresponds to a measurement error of 1/511, or about 0.2%. The voltage scale is from 0 to 10.22 V. Before beginning operation the voltmeter can be tested against known voltage calibrated on a standard cell. These tests show that the voltmeter consistently guarantees an error of 0.2%.

2. A measurement time reference system, consisting of a highly stable quartz oscillator with frequency dividers, generating pulses with a frequency from 1/8 to 8 cps. Consistent with this, the instrument can perform measurements at time intervals determined by the indicated frequencies: 1/8, 1/4, 1/2, 1, 2, 4, 8 cps. Furthermore, it is provided with the capability of measuring by an external trigger signal. The maximum frequency is limited at present by the speed of operation of the mechanical perforator.

3. A system for matching the operation of the perforator and digital voltmeter. This system distributes the numbers obtained in the measuring process to a 36-place output register, and translates a number in real time from this register to the perforator. In this way a denser compaction of data on the punched tape is possible, and the upper frequency limit of the measurements is raised.

4. A mechanical perforator for punching of the holes in the paper tape. The standard perforator provided as part of the Ural system is used. The perforator operates at two holes per second.

5. The measurement automation system, which enables all of the necessary peripheral operations to be executed automatically, including punching of the zone number ahead of the numerical material, change of zone number, insertion of a certain number of zeros between zones, fixed number entry. During operation of the system the control panel displays the measured number itself, the number of measurements, the number of the zone in which the numerical material is located, and certain other operational characteristics of the system. Moreover, the control panel has a set of keys for stopping the system after execution of the number of measurements indicated by depression of the keys. This all adds to the convenience of operation and elevates the level of automation such that the entire operation is reduced by the system to a matter of pushing the right button. As a result, the output of the perforator is a punched tape which, once wound, is completely ready for input to the machine. Consequently, human participation in the data recording process is reduced to a bare minimum.

The subsequent processing of the spectrum is also performed automatically by means of the Ural general purpose digital computer, which is controlled by a special program. The realization of a particular kind of spectrum processing, for example a definite technique for reduction to an ideal instrument, is now reduced to writing a suitable program. The capability of the machine to execute the required program is dictated by the capacity of the machine memory or, in the final analysis, by the machine time available for computation.

Hence it is possible to achieve practically total automation of the spectrum measurement process right down to the stage of obtaining the final results in digital form.

Three different programs have been written so far. Two of these perform two different methods of reduction and will be described in subsequent sections of this chapter. The third program performs statistical processing of the noise and is described in the author's dissertation [31], Chapter IV, which is devoted to an investigation of the characteristics of a typical infrared spectrophotometer.

§3. Reduction by the Fourier-Transform Method

The most general method of reduction to an ideal instrument, as we discovered in the Introduction, can be accomplished by means of Fourier transforms. However, because of the enormous difficulty of the computations

*The author would like at this point to express his sincere appreciation to B. E. Marchuk and the other members of the Ural computer group for their frequent assistance with our work.

involved in calculation of the Fourier transforms, this method has almost never been used. We attempted to use the Ural computer to carry out such a reduction. This required computations described by the formula

$$\varphi_a(\lambda) = \frac{1}{2\pi} \int_{-\infty}^{\infty} F(\omega) \, G(\omega) \, e^{-i\omega\lambda} \, d\omega. \tag{IV.1}$$

The program was written for $G(\omega)$ in the form

$$G(\omega) = \begin{cases} A^{-1}(\omega) & \omega \leqslant \omega_0, \\ 0 & \omega > \omega_0. \end{cases} \tag{IV.2}$$

It was assumed that ω_0 had been chosen beforehand according to Eq. (II.21). Consequently, to obtain the reduced contour of $\varphi_a(\lambda)$ it was necessary to perform calculations described by the following equation:

$$\varphi_a(\lambda) = \frac{1}{\pi} \int_0^{\omega_0} \frac{1}{A'^2(\omega) + A''^2(\omega)} \{(F'A' + F''A'')\cos\omega\lambda + (F''A' - F'A'')\sin\omega\lambda\} \, d\omega, \tag{IV.3}$$

which uses the notation

$$F(\omega) = F'(\omega) + iF''(\omega), \qquad A(\omega) = A'(\omega) + iA''(\omega).$$

The functions F', F", A', A" are calculated from the contours of $f(\lambda)$ and $a(\lambda)$ according to the equations

$$F'(\omega) = \int_{-\infty}^{\infty} f(\lambda)\cos\omega\lambda \, d\lambda = \int_0^{\infty} [f(\lambda) + f(-\lambda)]\cos\omega\lambda \, d\lambda,$$

$$F''(\omega) = \int_{-\infty}^{\infty} f(\lambda)\sin\omega\lambda \, d\lambda = \int_0^{\infty} [f(\lambda) - f(-\lambda)]\sin\omega\lambda \, d\lambda, \tag{IV.4}$$

and, analogously,

$$A'(\omega) = \int_0^{\infty} [a(\lambda) + a(-\lambda)]\cos\omega\lambda \, d\lambda,$$

$$A''(\omega) = \int_0^{\infty} [a(\lambda) - a(-\lambda)]\sin\omega\lambda \, d\lambda. \tag{IV.5}$$

To carry out the indicated calculations we wrote a program to execute the following operations:*

1. Input of numerical information, specifying the observed distribution $f(\lambda)$. In recording $f(\lambda)$ the range is chosen so that at the end points the spectrum will have segments which are parallel on the average to the horizontal axis. After input of $f(\lambda)$ ten points from each end are averaged, the mean height of the left and right wings is calculated, and a straight line is drawn through the end points of the recorded interval, the ordinate values of the straight lines at these points being set equal to the mean height of the corresponding wing. So drawn, this line comprises the background line $i_\varphi(\lambda)$. This background is subtracted from the observed distribution, and in subsequent processing the distribution without background is used:

---

*Henceforth we will omit the description of secondary computations, such as the number of points, summation check, parity check, etc.

$$f(\lambda) - i_\Phi(\lambda).$$

2. Computation of the Fourier transform of the resultant contour minus background and storage of the values of $F'(\omega)$ and $F''(\omega)$ in memory.

3. Input of numerical information, specifying the instrument function. As in the case of $f(\lambda)$, the background is substracted.

4. Computation of the Fourier transform of the instrument function $A'(\omega)$ and $A''(\omega)$ and the functions $\Phi'_a(\omega)$ and $\Phi''_a(\omega)$:

$$\Phi'_a(\omega) = \frac{F'A' + F''A''}{A'^2 + A''^2}, \quad \Phi''_a(\omega) = \frac{F''A' + F'A''}{A'^2 + A''^2}. \tag{IV.6}$$

The values of $\Phi'(\omega)$ and $\Phi''(\omega)$ are stored in the memory.

5. Computation of the inverse Fourier transform, which yields

$$\varphi'_a(\lambda) = \varphi_a(\lambda) + \varphi_a(-\lambda), \quad \varphi''_a(\lambda) = \varphi_a(\lambda) - \varphi_a(-\lambda). \tag{IV.7}$$

6. Computation of the final reduced distribution $\varphi_a(\lambda)$, addition of the background; printout of the resultant contour in tabulated form.

All three of the Fourier transforms in this program are computed with the same routine (Fourier integral computation subprogram) by substitution of the initial data. Computation of the integrals was programmed by the method of Simpson.

The compiled program was tested with simulation bands. The simulated observed contour was a contour of Gaussian configuration $f(\lambda) = \text{const} \cdot e^{-\lambda^2/2}$, $\delta \simeq 2.35$, the function $a(\lambda)$ also had a Gaussian form with $\sqrt{2}$ times the width, $\alpha \simeq 1.66$. The true contour in this case should also be Gaussian with a width equal to that of $a(\lambda)$, $\gamma = 1.66$. Figure 21 shows the true contour and the contour obtained by approximate reduction according to the compiled program. The Fourier transforms were cut off at $\omega_0\gamma = 16.6$. With cutoff at $\omega_0\gamma = 20.0$ the discrepancy between the reduced contour and the true one would have been difficult to detect in the figure. We see that these two contours are in rather good agreement. The greatest deviation is observed at the band peak and in the wings, where reduction induces small oscillations associated with the so-called Gibbs effect [92, 99], which occurs when the Fourier transform is cut off too sharply. It was found later, in the processing of Raman scattering lines with a nearly dispersion type configuration, that in this case the added oscillations may be very considerable. This is because $\Phi(\omega)$ has a relatively slowly decaying wing for lines with a dispersion configuration, which leads to large oscillations with sharp cutoff of the spectrum.

In this connection, it would be desirable to investigate further the conditions for reduction with another choice of $G(\omega)$ and to develop a new modification of the program to include the new form of $G(\omega)$. It would also be desirable to incorporate into the program a computation of the parameters for optimum $G(\omega)$, even if a rough approximation, for example according to the equations obtained for sharp cutoff.

Our program for reduction by the Fourier transformed method requires about one to two hours of machine time for the processing of a single band. If a new modification can be worked out to include the features indicated above, this time should increase even more. As we will see presently, the insufficient speed of the Ural computer at our disposal substantially limits the possibilities of using it in general for the solution of the complex problems of processing spectra. Consequently, it would be most desirable to have a faster machine for handling the program modification and for the solution of more complex problems in the processing of spectral data. At the present time a fair number of computers is available on which these problems can be solved with considerable success and in a short time. And there are no major difficulties involved in transferring the present programs to other machines.

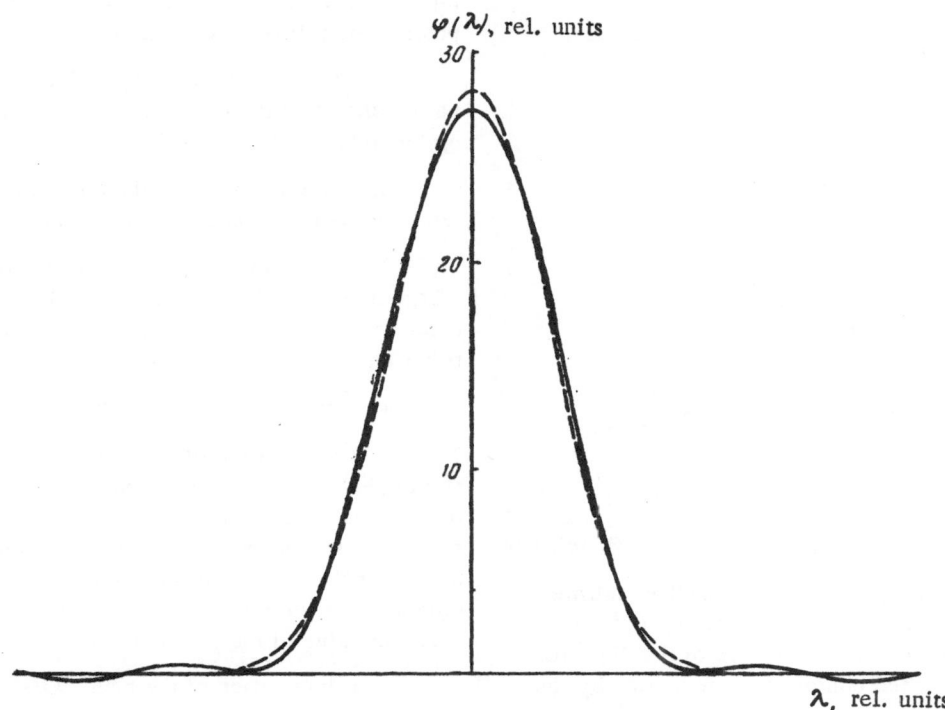

Fig. 21. Reduction by Fourier-transform method. The solid curve represents the contour obtained after approximate reduction; the broken curve is the true Gaussian contour.

§4. Rayleigh Method

In view of the fairly limited speed of the Ural computer, the time required for elimination of the instrument function according to the most general method using Fourier transforms is relatively long. For this reason the only possible application of this program seems to be in those situations when the band structure is complex and the instrument function is broad. As more rapid computers eventually become available all information should prove amenable to this method.

For the present we have developed a spectral processing program using Bracewell's refined version of Rayleigh's method [2, 14] for the processing of spectra recorded with a relatively narrow instrument function. This method is convenient for ordinary applications in that it takes advantage of Rayleigh's simple graphical construction, which greatly facilitates reduction. The reduction procedure itself amounts to the following: To provide the proper correction at any point λ_0 of the absorption curve a chord is drawn through points of the curve situated at a distance q in both directions from the λ axis at the point λ_0. At the point λ_0 the chord will be separated from the absorption curve by a segment h. A part of this segment βh then gives the desired correction at the point λ_0, displaced in the opposite direction from the chord. The quantity β depends on q and on the shape of the instrument function. The optimum value of q corresponding to the optimum total error, as shown in [20], is near $q = \alpha/\sqrt{2}$ (where α is the width of the instrument function). With this choice of q the value of β for a Gaussian instrument function is equal to $\frac{4}{11}$, for a triangular instrument function to $\frac{1}{3}$. This method provides a good approximation when the width of the instrument function $\alpha < \delta/2$.

The program written to execute this method has the following steps.

The system described in §2 of this chapter is used to record the investigated spectrum on punched tape. Once the record is made, the width of the instrument function α is keyed onto the same tape from the control panel. If an absorption spectrum is being recorded, additional reference points are entered, giving the values of total and zero absorption. Several spectra can be recorded in different zones (channels) of the same punched

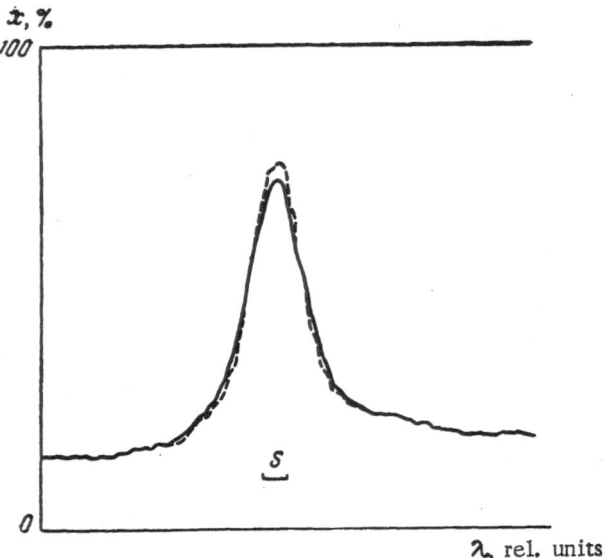

Fig. 22. Example of reduction by the Rayleigh method, for the benzene band $\nu = 1037$ cm^{-1}. The solid curve represents the record of the line, the broken curve is the contour after reduction by the Rayleigh method.

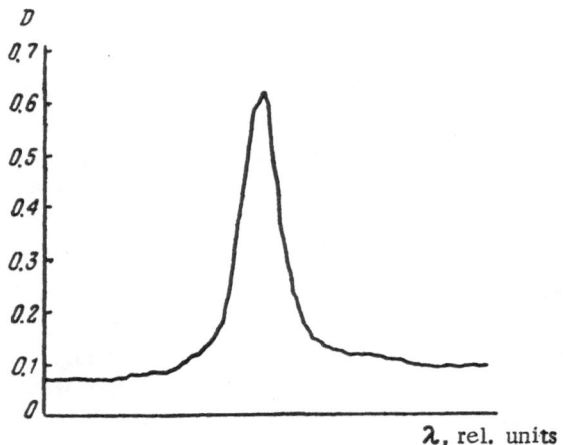

Fig. 23. Reduced contour of the benzene band $\nu = 1037$ cm^{-1}, transformed to the optical density.

tape. The punched tape is further processed in the computer. The following operations are executed:

1. Processing by the Rayleigh method. The interval q defined by the chord is automatically made nearly optimum [20, 59]: $q \simeq \alpha \sqrt{2}$.

2. Transformation to the optical density. The resultant contour is printed out in tabular form.

3. Averaging of the band wings, construction and subtraction of the background line. This is done just as described in §3. The parameters of the resultant distribution are then computed.

4. Computation of the total intensity.

5. Computation of the intensity or optical density at peak value. For this computation ten points of the contour distributed about the point of peak intensity are used; they are fitted by the method of least squares to a parabola, and the peak of this parabola is determined. This results in an effectively averaged, more accurate value of the peak intensity.

6. Computation of the band half-width γ_a, again by averaging of segments of the contour about points distributed about the half-power points, using the method of least squares; the contour in this is approximated by straight lines.

7. Computation of the form factors* r and K_f.

8. Computation of the record interval in band widths λ/γ_a. In the machine processing of spectra any of the routines can be dropped from the program. This means, for example, that a single program can be used to process both emission and absorption spectra, to obtain the parameters of the observed and the corrected band, as well as those of absorption and optical density bands.

Figures 22 and 23 illustrate the program just described in the processing of the benzene infrared band $\nu = 1037$ cm^{-1}; Figure 24 shows the approximation of the neighborhood of the band peak with a parabola. For this example the band $\nu = 1037$ cm^{-1} was recorded at a tray thickness d $\simeq 0.01$ mm, using the AIKS-F-4 spectrometer with an NaCl prism. The slit width was $S_1 = S_2 \simeq 0.250$ mm $= 5.36$ cm^{-1}, the scan rate was $\nu = 0.122$ cm^{-1}/sec, the time constant was $\tau = 2$ sec. The observed contour was recorded at 230 points with a measurement frequency $f_{meas} = 1/4$ cps.

*The form factor r $= \pi \delta I_0/2 I_\infty$ was introduced by M. M. Sushchinskii [9, 100] as a characteristic of the shape of the Raman optical scattering lines. We also use the form factor $K_f = s_\delta/s_{n\delta}$, where s_δ is the area of the band contained between the half-power points (half-width δ), $s_{n\delta}$ is the area under the wings from the half-width to $n\delta$.

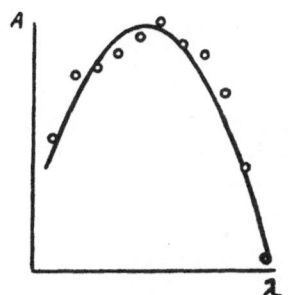

Fig. 24. Peak absorption, computed by the method of least squares, with the band peak approximated by a parabola.

The absolute figures for the band parameters are not given, since in these measurements, which were undertaken to check out the program, the distance between reference point was not determined exactly, i.e., the scale on the wavelength axis was not measured with sufficient precision. The total time to process the spectrum on the Ural computer was on the order of 15 min.

CONCLUSION

Thus we have seen that for the practical application of the above analysis of instrument distortions and methods for their correction it is necessary to understand the properties of the spectra investigated, as well as the properties of the spectral instrumentation, specifically their instrument function and accuracy characteristics. Inasmuch as such data are almost totally lacking, measurements have been made for the purpose of bridging this gap [31, 101]. A brief summary of the results obtained is given below.

Measurements performed with high-resolution instruments have shown that the shape of individual infrared absorption bands is clearly very close to a dispersion configuration. It is necessary, however, to bear in mind that individual bands, undistorted by the superposition of neighboring bands, are seldom actually encountered in isolation. The table below lists data for a number of bands in the 8 to 11 μ region of the spectrum. It gives the frequency ν and wavelength λ of the investigated bands, as well as the thickness of the experimental tray d. The last columns of the table give the measured values of the peak absorption x_0, peak optical density D_m, and width of the bands γ_x (absorption half-width) and γ_D (optical density half-width).

Included among these bands are two whose shape is very nearly a dispersion configuration. This renders them very amenable to evaluation of the width of the instrument function for medium dispersion instruments[31].

An investigation was made of the properties of a typical infrared instrument [31]. A twin-wave infrared spectrophotometer designed at the Physics Institute (FIAN) with a type IKS-11 monochromator was studied. Measurements made according to two procedures (by means of an instrument with high resolving power and by means of the abovementioned dispersion band) showed that the instrument function of the IKS-11 monochromator with NaCl prism is almost Gaussian in shape, while its width in the working range is essentially determined only by the width of the monochromator slits.

Furthermore, the accuracy characteristics of the same instrument were also investigated. Processing of the measurements on the Ural digital computer produced the following results. The probability distribution

Substance	ν, cm^{-1}	λ, μ	d, μ	x_0, % background	D_m	γ_x, cm^{-1}	γ_D, cm^{-1}	Shape
Solid naphthalene . . .	1008	9.91	21.5	76	1.43	2.3	1.6*	Side bands
Pyridine	1030.7	9.702	18	86	1.93	6.8	4.6	Weak side band
Pyridine	1147.4	8.715	18	54	0.76	8.4	6.9	Dispersion
Pyridine	1217.6	8.213	18	53	0.75	8.8	7.1	Side band
Cyclohexane	903	11.06	40	74	1.36	6.2	4.5†	Dispersion
Benzene	1036.8	9.645	10	68	1.13	12.5	8.9	Side band
Acetonitrile	918	10.90	20	61	0.95	9.0	7.9	Asymmetrical

*In [102] γ_D = 1.8 cm^{-1}

†In [102] γ_D = 5.0 cm^{-1}

for the noise level may be considered Gaussian correct to be the measurement error. The noise level is fairly accurately described by the formula

$$\overline{\Delta l^2}/I_0^2 = E/\tau S_1^2 S_2^2,$$

and with equal slit widths $S_1 = S_2 = S$ by the formula

$$\overline{\Delta l^2}/I_0^2 = E/\tau S^4.$$

For the variable E with $\lambda = 10.65 \, \mu$ the measurements gave $E = 1.7 \cdot 10^{-6} \, \text{sec} \cdot \text{mm}^4$. The mechanical error was also measured. For our instrument it came to about 0.5%.

These characteristics may be regarded as approximate estimates of the properties inherent in the most generally used class of infrared spectrometers: medium dispersion instruments.

The present work was begun on the initiative and under the direction of Grigorii Samuilovich Landsberg. His advice and steadfast support rendered the author inestimable aid.

The author takes this opportunity to express his gratitude to Prof. P. A. Bazhulin for his constant interest in the work, to V. I. Malyshev and S. G. Rautian for their assitance and valuable discussions.

LITERATURE CITED

1. J. W. Strutt (Lord Rayleigh), The Wave Theory of Light [Russian translation],Gostekhizdat (1940).
2. Rayleigh, Sci. Papers 1: 135 (1871).
3. S. G. Rautian, Usp. Fiz. Nauk 66: 475 (1958).
4. A. A. Kharkevich, Spectra and Analysis [in Russian], Gostekhizdat (1957).
5. A Eddington, Monthly Notices Roy. Astron. Soc., 73: 359 (1913).
6. T. M. Young and A. C. Hurdy, J. Opt. Soc. Amer. 39: 265 (1949).
7. B. Van der Pol and H. Bremmer, Operational Calculus [Russian translation] IL (1952).
8. I. V. Peisakhson, Izv. Akad. Nauk SSSR, Ser. Fiz. 18: 680 (1954).
9. M. M. Sushchinskii, Tr. Fiz. Inst. Akad. Nauk 12: 54 (1960).
10. F. Paschen, Wied. Ann. 60: 712 (1897).
11. C. Runge, Z. Math. 42: 205 (1897).
12. E. P. Hyde, Astrophys. J. 35: 237 (1912).
13. W. H. Eberhardt, J. Opt. Soc. Amer. 40: 172 (1950).
14. R. N. Bracewell, J. Opt. Soc. Amer. 45: 873 (1955).
15. H. C. Van de Hulst, Bull. Astron. Inst. Neth. 9: 225 (1941).
16. S. G. Mikhlin, Integral Equations, Gostekhizdat (1947).
17. H. C. Burger and P. H. van Cittert, Z. Physik 79: 722 (1932).
18. V. M. Chulanovskii and A. V. Timoreva, Izv. Akad Nauk SSSR, Ser. Fiz. 11: 376 (1947).
19. G. S. Gorelik, Dokl. Akad. Nauk SSSR 83: 549 (1952).
20. C. G. Rautian, Masters Thesis, Fiz. Inst. Akad. Nauk (1957).
21. C. G. Rautian, Dokl. Akad. Nauk SSSR 109: 743 (1956).
22. L. A. Khalfin, Dokl. Akad. Nauk SSSR 122: 1007 (1958).
23. B. S. Tsybakov and V. P. Yakovlev, Izv. Vysschikh. Uchebn. Zavedenii, radiofiz. 1: 98 (1958).
24. G. Toraldo di Francia, Nuovo Cimento Suppl. 9: 426 (1952).
25. L. A. Khalfin, Proceedings of the All-Union Conference on Probability Theory and Mathematical Statistics, Yerevan, Armenian SSR (1960), pp. 187-205.
26. L. A. Khalfin, (in press).
27. V. M. Bunimovich, Fluctuation Processes in Radio Receiving Systems, Izd. "Sov. radio" (1951).
28. V. M. Chulanovskii, Introduction to Molecular Spectral Analysis, Gostekhizdat (1961).
29. S. Brodersen, J. Opt. Soc. Amer. 43: 877 (1953).
30. D. Z. Robinson, Anal. Chem. 23: 273 (1951).
31. G. G. Petrash, Author's abstract of Masters Thesis, Moscow, MFTI (1961).
32. G. G. Petrash and S. G. Rautian, Inzh.-Fiz. Zh. 1(7): 161 (1958).
33. G. G. Petrash and S. G. Rautian, Materials of the Tenth All-Union Conference on Spectroscopy, L'vov (1957), p. 102.
34. G. G. Petrash, Opt. i Spektroskopiya 6: 793 (1959).
35. G. G. Petrash, Optik aller Wellenlängen, Berlin (1959).
36. G. G. Petrash, Opt. i Spektroskopiya 8: 122 (1960).
37. G. G. Petrash, Opt. i Spektroskopiya 9: 423 (1960).
38. M. Schubert, Exptl. Tech. Physik 6: 203 (1958).

39. W. Rohman and M. Schubert, Optik aller Wellenlängen, Berlin (1959).

40. L. A. Gribov, Pribory i Tekhn. Eksperim. (2): 65 (1958); (3): 102 (1959).

41. I. V. Peisakhson, Opt. i Spektroskopiya 8: 116 (1960).

42. V. M. Chulanovskii, I. V. Peisakhson, and D. N. Shchepkin, Opt. i Spektroskopiya 7: 763 (1959); 8: 57 (1960).

43. A. I. Tudorovskii, Theory of Optical Instruments, Chapt. II, Izd. AN SSSR (1952).

44. H. Moser, Optik 12: 362 (1955).

45. H. Frieser, Phot. Korr. 92: 183 (1956).

46. V. V. Solodovnikov, Introduction to the Statistical Dynamics of Automatic Control Systems, Gostekhizdat (1952).

47. A. S. Toporets, Monochromators, Gostekhizdat (1956).

48. V. N. Smirnov, Vestn. Mosk. Univ., Ser. Mat., Mekhan., Fiz. (1): 61 (1959).

49. S. Brodersen, J. Opt. Soc. Amer. 43: 1216 (1953).

50. I. S. Abramson and A. N. Mogilevskii, Izv. Akad. Nauk SSSR, Ser. Fiz. 19: 49 (1955).

51. A. V. Ioganson, Masters Thesis, Moscow, All-Un on Scientific Research Institute of the Petroleum Industry (1954).

52. A. R. Philpotts, W. Thain, and P. G. Smith, Anal. Chem. 23: 268 (1951).

53. S. Brodersen, J. Opt. Soc. Amer. 44: 22 (1954).

54. O. D. Dmitrievskii and V. A. Nikitin, Zh. Opt.-Mekh. Prom. (4): 9 (1957); (2): 26 (1958); (6): 25 (1958).

55. O. D. Dmitrievskii, B. S. Neporent, and V. A. Nikitin, Usp. Fiz. Nauk 64: 447 (1958).

56. V. A. Nikitin, Opt. i Spektroskopiya 4: 523 (1958).

57. E. F. Daly and G. B. Sutherland, Proc. Phys. Soc. A62: 205 (1949).

58. R. Zbinden and E. Baldinger, Helv. Phys. Acta 26: 111 (1953).

59. G. G. Petrash and S. G. Rautian, Inzh.-Fiz. Zh. 1(11): 80 (1958).

60. S. G. Rautian and G. G. Petrash, Materials of the Tenth All-Union Conference on Spectroscopy, L'vov (1957), p. 107.

61. L. S. Roess, Rev. Sci. Instru. 16: 172 (1945).

62. R. C. Jones, J. Opt. Soc. Amer. 39: 327, 344 (1949).

63. M. N. Markov, Pribory i Tekhn. Eksperim. (3): 70 (1956).

64. M. N. Markov, Dokl. Akad. Nauk SSSR 108: 428 (1956).

65. R. A. Smith, F. E. Jones, and R. P. Chasmar, The Detection and Measurement of Infrared Radiation [Russian translation] IL (1959).

66. A. M. Bonch-Bruevich and Ya. A. Imas, Zh. Techn. Fiz. 25: 2565 (1955).

67. A. M. Bonch-Bruevich and Ya. A. Imas, Izv. Akad Nauk SSSR, Ser. Fiz. 19: 54 (1955).

68. Ya. A. Imas, Pribory i Tekhn. Eksperim. (6): 100 (1958).

69. J. B. Willis, Australian J. Sci. Res. A4: 172 (1951).

70. G. G. Petrash, Inzh.-Fiz. Zh. 1(9): 94 (1958).

71. G. G. Petrash, Physical Problems of Spectroscopy, Vol. 1, Izd. AN SSSR (1963), p. 62.

72. B. P. Ramsay, E. L. Cleveland, and O. T. Coppius, J. Opt. Soc. Amer. 31: 26 (1941).

73. A. Schuster, Astrophys. J. 21: 197 (1905); A Schuster, Theory of Optics [Russian translation] ONTI-GTTI (1935).

74. S. É. Frish, Techniques of Spectroscopy, Izd. LGU (1936).

75. S. Tolanskii, High-Resolution Spectroscopy [Russian translation] IL (1955).

76. K. Meissner, J. Opt. Soc. Amer. 31: 405 (1941); 32: 185 (1942).

77. S. S. Mitra, J. Sci. Ind. Res. 14B: 303 (1955).

78. C. M. Sparrow, Astrophys. J. 44: 76 (1916).

79. F. A. Korolev, High-Resolution Spectroscopy, Gostekhizdat (1953).

80. W. V. Houston, Phys. Rev. 24: 478 (1927).

81. A. Buxton, Phil. Mag. 23: 440 (1937).

82. K. C. Chaturvedi and M. S. Sodha, Indian J. Phys. 30: 491 (1956).

83. M. S. Sodha, Indian J. Phys. 28: 141 (1954).

84. K. C. Chaturvedi and M. S. Sodha, Indian J. Phys. 30: 543 (1956).

85. K. C. Chaturvedi and M. S. Sodha, Indian J. Phys. 30: 599 (1956).

86. K. C. Chaturvedi, Optik 17: 34 (1960).

87. A. C. Emsly and G. W. King, J. Opt. Soc. Amer. 41: 405 (1951); 43: 658 (1953).

88. S. S. Mitra and M. S. Sodha, Optik 14: 277 (1957).

89. I. V. Peisakhson, Opt. i Spektroskopiya 5: 209 (1958).

90. A. A. Michelson, Light Waves and Their Uses [Russian translation] Gostekhizdat (1934).

91. P. Fellgett, J. Phys. Radium 19: 197 (1958).

92. J. Connes, J. Phys. Radium 19: 197 (1958).

93. W. Weizel and G. Meister, Z. Physik 144: 177 (1956).

94. T. Williams, J. Opt. Soc. Amer. 49: 1138 (1959).

95. V. I. Malyshev, M. N. Markov, and A. A. Shubin, Dokl. Akad. Nauk SSSR 86: 273 (1952); Izv. Akad. Nauk SSSR, Ser. Fiz. 17: 654 (1953).

96. D. A. Ramsay, J. Amer. Chem. Soc. 74: 72 (1952).

97. A. I. Kitov and N. A. Krinitskii, Electronic Digital Computers and Programming, Fizmatgiz (1959).

98. B. E. Marchuk, Peredovoi Nauch.-Tekhn. i Proizv. Opyt (9) (1961).

99. J. Strong and G. Vanasse, J. Opt. Soc. Amer, 49: 844 (1959).

100. P. A. Bazhulin and M. M. Sushchinskii, Usp. Fiz. Nauk 68: 135 (1959).

101. G. G. Petrash, Opt. i Spektroskopiya 9: 121 (1960).

102. R. A. Russel and H. W. Thompson, Spectrochim. Acta 9: 133 (1957).

STUDY OF THE EFFECT OF TEMPERATURE
ON THE RAMAN SPECTRA OF SUBSTANCES
IN VARIOUS STATES OF AGGREGATION

A. I. Sokolovskaya

INTRODUCTION *

The Raman scattering of light (Raman effect) gives especially valuable information in studying the structure of materials, since this phenomenon is closely connected with the configurations and dynamics of the electrons and nuclei in the molecule. After the discovery of the Raman effect, a large number of investigations devoted to the study of this phenomenon were carried out, as a result of which the theory of the Raman effect for noninteracting molecules has at the present time been fully developed [1-5]. However, the overwhelming majority of experimental investigations into Raman spectra have been carried out in the liquid phase, in view of their low intensity. In this case the light-scattering molecule interacts with surrounding molecules, and the experimentally observed spectra must reflect the properties of a system of interacting molecules. This fact has often not been considered by experimenters, and results obtained from investigations in the condensed phase of the substance have been compared with the data of a theory which was developed for the case of noninteracting molecules.

Recently, there has been an increase in the number of publications indicating that the total intensity and also the distribution of energy in the contour of the Raman spectral lines for the condensed phase of matter may be substantially distorted under the influence of intermolecular forces. However, up to the present, the effect of molecular interactions on Raman spectra has been one of the least studied questions of molecular spectroscopy. The fact that, for a long time, there has been no practical study of the effect of temperature on the integral intensity and line shape of the Raman effect also bears witness to this.

Nevertheless, a study of the effect of temperature on these parameters of the Raman line is of great scientific and practical interest. In studying the temperature dependence of the intensity and shape of Raman lines for gases at low pressures, we may obtain information on the probabilities of transitions, and on the character of processes producing the attenuation of the characteristic oscillations of noninteracting molecules. A study of the intensity and shape of the lines in the Raman effect for various temperatures in the condensed phase, and also their behavior on passing from gas to liquid and from liquid to solid, may provide an explanation of such leading problems as the nature of the intermolecular action, and the structure of liquids and solids. A series of recent publications [6-11] has shown that the Raman line widths may serve as a source of information on the kinetics of relaxation processes in condensed states of matter. An exact knowledge of the laws of change in the intensity and shape of the Raman lines with temperature is necessary for a valid approach to the interpretation of experimental data on the spectra of molecules in liquids and solids. Finally, the temperature dependence of the intensity and breadth of the lines must be known in carrying out quantitative molecular analysis of various hydrocarbon mixtures from Raman spectra.

The present paper is devoted to a study of the effect of temperature on the Raman spectra of substances in various states of aggregation.

In order to obtain the most complete picture of the effect of temperature on the spectrum of a substance, we must simultaneously study the temperature dependence of the integral intensity, breadth, and shape of the Raman lines. With a view to setting up limits to the applicability of the elementary theory of the temperature

* Dissertation in partial fulfillment of the requirements for the degree of Candidate in Physical and Mathematical Sciences. Defended at the Physical Institute, Academy of Sciences of the Belorussian SSR. Research Director: Doctor of Physical and Mathematical Sciences Professor P. A. Bazhulin.

dependence of Raman line intensities [1-3], we thought it appropriate to carry out measurements of the spectra of substances in the liquid and gaseous states both in the Stokes and anti-Stokes parts of the spectrum.

In order to explain the nature of the effect of intermolecular action on the intensity of spectra, it was of interest to study the intensity of lines on passing from liquid to vapor, and also in solutions at various temperatures.

CHAPTER I

TEMPERATURE DEPENDENCE
OF THE INTENSITY OF RAMAN SCATTERING LINES

Soon after the discovery of the Raman phenomenon, I. E. Tamm, G. S. Landsberg, and L. I. Mandelstam [1] showed on the basis of quantum theory, allowing for the statistical distribution of molecules with respect to the vibrational levels, that the intensity of the Stokes lines was

$$I_{St} \sim \frac{1}{1 - e^{-\frac{h\nu_j}{kT}}},$$ (1)

and that of the anti-Stokes lines

$$I_{ASt} \sim \frac{1}{e^{\frac{h\nu_j}{kT}} - 1},$$ (2)

where ν_j is the vibration frequency. The ratio of the intensitites of Stokes and anti-Stokes lines in a region far from resonance is given by the formula

$$\frac{I_{St}}{I_{ASt}} = \left(\frac{\nu - \nu_j}{\nu + \nu_j} \right)^4 e^{\frac{h\nu_j}{kT}},$$ (3)

where ν is the frequency of the exciting light. In Raman frequency regions close to the electronic absorption bands, formula (3) is invalid, since the dependence of the scattering moment on the frequency must be taken into account [2].

The relations given above were derived for noninteracting molecules. Yet, ever since the discovery of the Raman effect, investigations into the temperature dependence of the Raman line intensity have been conducted on the liquid and even solid phases of matter, the results then being compared with relations (1) to (3). Meanwhile, the nature of the spectral distortion under the influence of intermolecular forces remained unknown.

First of all in 1929 Krishnan [12], and then Dadien and Kohlrausch [13] observed in the spectrum of liquid carbon tetrachloride a weakening in the intensity of the Stokes and an increase in the intensity of the anti-Stokes lines on raising the temperature. A similar phenomenon was noted by Brickwede and Peters [14] in 1929 in studying the spectrum of crystalline quartz at temperatures of −180 and +20°C. The results of these investigations agreed with the semiclassical explanation of the Raman effect existing at the time. In 1930, however, I. E. Tamm, G. S. Landsberg, and L. I. Mandelstam [1] showed that the data of the work in question did not agree with quantum theory, which allowed for the statistical distribution of the molecules with respect to the energy levels. According to the new theory, the intensity of the Stokes and anti-Stokes components should increase with temperature [see formulas (1) and (2)], and the ratio of these intensities should at any temperature obey relation (3). G. S. Landsberg and L. I. Mandelstam showed experimentally that the intensity of the Stokes and anti-Stokes satellites $\Delta\nu = 465$ cm^{-1} in the spectrum of crystalline quartz rose on increasing the temperature from 20 to 500°C, just as in formulas (1) and (2).

The ratio of the Stokes and anti-Stokes intensities coincides with that calculated from (3). The deviation of the results published by Brickwede and Peters from elementary theory was explained by the authors from the fact that, in the investigations in question, only the intensity at the line maximum was measured, rather than the integral intensity. The temperature variation of this quantity is more a measure of the change in line shape than the change in the intensity of the scattered light.

Later, Ornstein and Went [15, 16], Anantakrishnan [17], and Venkatesvarlu [18-24] published a series of papers devoted to studying the effect of temperature on the spectra of quartz, calcite, mercurous chloride, and sodium nitrate crystals, and several liquids. For all these substances there was a fall in the intensity of the Stokes components as temperature increased, thus contradicting the conclusions of the elementary theory. Furthermore, the ratio of the Stokes and anti-Stokes intensities coincided with the theoretical value. Anantakrishnan and Venkatesvarlu considered that the cause of the observed deviation from elementary theory of the temperature dependence of the Stokes lines in Raman scattering might lie in a change of the dynamics of the intramolecular vibrations of the scattering molecules under the influence of temperature. The validity of this assertion, however, could not be established exactly, owing to the absence of experimental measurements in the gaseous phase.

In 1950, L. M. Fishkova [25] carried out a temperature investigation of the intensities of the Stokes lines in certain liquids (carbon tetrachloride, benzene, chlorobenzene) by a photoelectric method. The intensity was measured at the line maximum with monochromator slit width from 35 to 42 cm^{-1}. For the liquids studied there was a sharp fall in the intensities of all the Stokes lines (by roughly two to three times) on raising the temperature from -15 to $+70$°C. The author suggested that an explanation might well be found "by taking account of associative formations in the liquid, the number of which (and hence the light scattering) may change with temperature" [25]. This proposition may indeed be correct, but requires more detailed substantiation.

An attempt to explain the fall in Raman line intensities in liquids on raising the temperature was made by Roesler [26]. He showed theoretically that, on increasing the temperature, a considerable part of the energy may pass into the wings of the line (collision broadening). In view of this, the author considered, the observed anomaly could be explained by a failure on the part of experimentalists to make allowance for the low-intensity wings. Without touching on the causes of the line broadening, we may say that this suggestion has a certain basis, but it lacks experimental proof.

The most systematic investigation of the temperature dependence of intensity in the Stokes part of the Raman spectrum for a number of liquids was undertaken by Ya. S. Bobovich and colleagues [27-29]. Intensity measurements were made at the line maximum with wide slits in the spectral system over a temperature range 14 to 55°C. Analysis of the experimental data obtained led the author to the following conclusions: The intensity of the Stokes lines falls on raising the temperature; the temperature dependence of dipolar liquids is greater than that of nondipolar liquids; the intensities of lines corresponding to deformation oscillations of the molecules are the most susceptible to temperature changes; the intensities of valence vibrations, especially those of double bonds, change little on altering the temperature of the substance. In [29, 30] Ya. S. Bobovich considered the proposition that the observed phenomenon was caused by the combined effect of chemical bond properties and intermolecular interactions.

The review just given roughly indicates the state of the problem as our own investigation begins. In view of the growth of interest in this problem, there have recently been quite a number of papers on the temperature dependence of Raman scattering as well as of infrared absorption. The results of these investigations will be discussed in the course of setting out our own experimental data.

As we see from the literary review, the majority of authors comparing experimental results with theory have started from the tacit assumption that the theory developed for an isolated molecule is applicable to liquids and solids. The validity of this assertion could not be completely established at the moment of starting the present work owing to the lack of systematic studies of the Stokes and anti-Stokes spectral regions simultaneously. Study of the effect of temperature on the intensity of the Stokes and anti-Stokes parts of the spectrum and on the ratio of the Stokes and anti-Stokes component intensities has borne a somewhat random character, and as a rule has been carried out on a small number of substances (carbon tetrachloride, quartz). A more systematic investigation has been carried out over a small temperature range ($\approx 20°C$) in the Stokes region. It should be noted that measurements of the intensity ratio of the Stokes and anti-Stokes lines is most important, since this determines the applicability of statistical theory to the intensity of spectra in the condensed phases of the substance. The possibility of such a proof of the theory constitutes the advantage of studying the problem in the Raman spectra in comparison with infrared absorption spectra.

The complete absence of measurements of the temperature dependence of intensity in gases and vapors at low pressures (for which the elementary theory, in essence, was developed) prevents us from checking the validity of the theory for the case of weak intermolecular interactions, and establishing the cause of the anomalous behavior of the Raman line intensities for matter in a condensed phase.

It should be noted that, in all investigations devoted to the effect of temperature on the Raman spectra, only one parameter, the intensity, has been considered, whereas the change in line shape has remained unknown. The results of a number of investigations [31-37] show that, as a rule, changes in intensity and line shape are interconnected. Hence study of line shape at various temperatures could well give additional information on the mechanism of processes causing the anomalous temperature dependence of Raman line intensities for liquids and solids.

As we have already mentioned, the majority of investigations have revealed a deviation between experimental data and the conclusions of elementary theory. However, the changes in line intensity with temperature obtained by different authors differ not only in magnitude but in a number of cases in sign also.

In view of the fact that the problem of the effect of temperature on Raman spectra is of fundamental scientific and practical interest, we considered it appropriate to undertake an investigation into the temperature dependence of Raman line intensities both in the Stokes and anti-Stokes regions of the spectrum, and more deeply to study the connection between this effect and the molecular structure and dipole moment, as well as the type of vibration made by the atoms in the molecule and other characteristics of the material. In order to obtain fairly full information on the processes giving rise to the phenomenon under consideration, we considered it necessary to study the temperature dependence of the Raman line shapes as well as their intensities. A study of the line intensities on passing from the condensed phase of the material into the vapor, and also the temperature dependence of line intensities in solutions, will enable us to elucidate the role of the intermolecular interactions and their nature in the observed phenomenon. Since the Raman line parameters studied may be considerably distorted by the apparatus, we shall be paying special attention to methodical questions connected with their measurement.

CHAPTER II

METHODS OF MEASURING THE INTEGRAL INTENSITY
AND SHAPE OF RAMAN EFFECT LINES

§1   Determination of the True Integral Intensity and Shape of Raman Lines

The experimentally obtained integral intensity and shape of a line may differ substantially from the truth. This is due to the influence of distorting factors associated with specific characteristics of the Raman spectra, experimental conditions, and the measuring system, on the quantities being measured. As we know, Raman lines are quite weak and are usually superimposed on a continuous background. On changing the temperature of the substance, and also during phase transformations, the relative intensities of the line under consideration and the background may change. In this respect, one of the difficulties inherent in studying Raman spectra lies in taking proper account of the low-intensity wings (flanks) of the line.

In the present investigation, in studying the temperature dependence of Raman spectral intensities, we selected mainly nonoverlapping lines situated in a background with intensity small (5 to 6%) compared with that of the line in question. The integration range in this case covers from six to eight line widths. However, even in very unfavorable conditions (integration range three to four widths), errors associated with the fact that the wings were not taken into account played only a small part in our measurements, since in carrying out the integration we cut off the same part of the contour for all the lines being compared. The integral intensity was determined by measuring the area bounded by the line contour at the same number of widths. In this it was important that the lines in question were approximated by the same analytical curves.*

In cases in which the intensities of lines differing in shape were being compared, a correction for the "wings" was introduced by means of a "form" factor, using a method proposed by M. M. Sushchinskii [37]. Moreover, when the temperature and state of aggregation of a substance change, there is also a change in the refractive index, and this may have an effect on the scattered light flux recorded.

The whole array of effects connected with variations in the refractive index of a liquid reduces to the following factors:

1) A change in the coefficient of reflection of the exciting radiation at the boundary between the liquid and the glass of the containing vessel;

2) A change in the configuration of the exciting rays inside the vessel;

3) A change in the brightness of the scattered light on passing out of the vessel;

4) A change in the intensity of the Raman scattering light reflected from the walls of the vessel;

5) A change in the conditions of imaging the vessel at the spectrograph slit.

Let us first of all consider the factors susceptible to theoretical analysis.

---

*For the overwhelming majority of substances investigated in the present paper, the line shapes were approximated by a dispersion curve, the form of the lines not changing with temperature. This point will be treated in more detail later.

We accounted for the change in brightness of the scattered light on passing out of the vessel as in[38, 39], by multiplying the observed Raman line intensities by $n^2_{\lambda liq}$ ($n_{\lambda liq}$ = refractive index at the wavelength of the scattered light).

The effect of a change in the intensity of the Raman scattered light reflected from the walls of the vessel was completely eliminated in our apparatus. To this end, we used the method of adjusting the vessel proposed by S. G. Rautian [40]. In his calculations it was assumed that: 1) the height of the homogeneously illuminated part of the slit was a maximum; 2) the collimator was filled from every point of the slit; 3) the exciting flux scattered at inhomogeneities in the vessel walls and the Raman light reflected from the walls did not fall into the spectral system. With suitable adjustment, the condenser projects the entrance slit of the spectral system on the rear part of the bottom of the vessel, and the rim of the collimator objective of the spectrograph on the forward part of the bottom of the vessel. The size of the corresponding images is less than the diameter of the vessel. Moreover the maximum height of the homogeneously illuminated part of the slit equals

$$h_m = \frac{4r_1 r_2}{l} \frac{f}{a} n_{liq}. \tag{4}$$

where r_1 and r_2 are the radii of the diaphragms at the rear and forward planes of the vessel, l is the geometric length of the vessel, and f and a are the focal length and diameter of the collimator objective respectively.

The focal length F and minimum relative aperture D/F of the condenser lens are determined in the following way:

$$F = 2r_2 \frac{f}{a} \frac{1}{1 - (2r_2/a)^2 (f/l) n_{liq}}, \tag{5}$$

$$\frac{D}{F} = \frac{a + h_m}{f} + 2n_{liq} \frac{r_1 + r_2}{l}. \tag{6}$$

The distance of the vessel from the condenser x and that of the condenser from the slit y are

$$x = F(1 + 2r_2/a), \quad y = F(1 + h_m/2r_1). \tag{7}$$

In calculating the parameters of the condenser and its position for photoelectric recording on the DFS-4, it was considered that the distance between the diffraction grating and the collimator mirror was comparable with the focal length of the collimator. In this case, in order to use the whole light flux falling on the monochromator, an image of the diffraction grating in the collimator objective was projected on to the front of the bottom of the vessel.

It should be noted that with uniform illumination of the sample, using the given adjustment, the effects of changing the conditions of the vessel on the spectrograph slit (fifth factor) on the intensity of the scattered light are also removed. Under these conditions, the illumination E of the line is determined only by the geometrical length of the vessel and the parameters of the spectral system [40]

$$E = \frac{S}{n^2_{\lambda liq}} l\Omega, \tag{8}$$

where S is the volume brightness of the sample, l the length of the vessel, and Ω the solid angle of the collimator objective.

If the sample is illuminated inhomogeneously, then this adjustment does not completely eliminate the effect of the imaging conditions of the vessel, although it reduces this to a minimum. The effects of this factor will be considered a little later together with the second factor (change in the ray paths of the exciting light inside the vessel).

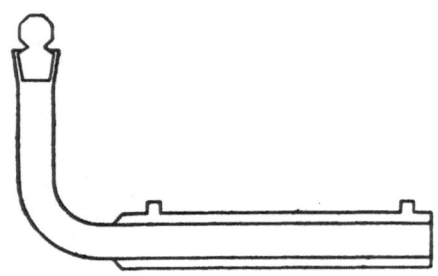

Fig. 1. Vessel with double walls for measuring the reflected part of the exciting radiation.

Thus only the third, fourth, and (partly) the fifth factors are immediately subject to calculation. The effects of the remainder (losses in reflection of the exciting light at the boundary between the liquid and the glass of the vessel and changes in the configuration of the exciting rays inside the vessel) may change from one apparatus to another, and exact allowance for these factors is practically impossible. This led the authors of [41, 42] to the conclusion that the only possibility of avoiding systematic errors lay in the calibration of the apparatus with a standard sample. These authors propose to vary the refractive index of the scattering liquid by changing either its temperature or the concentration of the solution, and to carry out a calibration using the known temperature or concentration dependence of the line intensities. But in order to ascertain this kind of dependency it is necessary to know just the very effect of the refractive index of the sample. The recommendation of [41, 42] thus leads to an argument in a circle. In order to ascertain the effect of reflection losses on the intensity measured, we used one of the possible arrangements in which the scattering substance was not varied [43].

In order to measure the reflected part of the exciting radiation, the vessel containing the scattering liquid was placed in a glass sleeve (Fig. 1); into the gap formed were poured in turn liquids with different refractive indices. The resulting two surfaces of separation (glass—liquid and liquid—glass) imitated the one surface existing in ordinary measurements and increased the coefficient of reflection by roughly a factor of two. It is important that in our experiments the scattering liquid and the vessel were not changed. Hence the coefficient of reflection at the glass—air and scattering liquid—glass boundaries remained constant. With sufficiently thin walls and narrow gap there was no need to fear changes in the paths of the rays inside the vessel, and we could therefore study the change in reflection of the incident rays in pure form.

In our vessel the wall thickness of the supplementary tube was 0.7 mm and the gap 0.5 mm. The refractive index of the vessel walls was $n_w = 1.52$. As scattering liquid we used toluene. The gap was filled successively with carbon disulfide (n = 1.67), toluene (n = 1.52), and ethyl alcohol (n = 1.36). Then the intensities of the lines of the scattering substance were measured as a function of the refractive index of the liquid filling the gap. Experiments were also made with the gap empty (n = 1). The measurements were made photographically on the HUET B-III spectrograph and also photoelectrically on the DFS-4. For photographic recording we used fine-grained "blau rapid" plates. The photoplate simultaneously recorded the spectrum of the scattering substance and that of the mercury lamp, which enabled us to eliminate oscillations in the intensity of the exciting radiation. The results of the measurements were averaged and are given in Table 1. This shows the intensities I of the Raman lines of toluene with the gap filled by various liquids. The intensity for a gap filled with toluene is taken as unity. Each value of I was obtained by averaging approximately 60 measurements. The mean square scatter of the measurements was around 1.5%. The mean deviation from the quantity measured was approximately 5%.

As we see from Table 1, the change in the incident light flux for real liquids practically lies within the limits of measuring error, and in the presence of only one surface of separation is considerably less than experimental error. Hence the effect of losses of exciting light by reflection in the sample on the Raman intensity may in our case be neglected. Naturally a different picture may arise under other conditions.

It should be noted that in the paper by Bernstein and Allen [42] the correction due to losses in reflection reaches quite high values. In particular, the authors asserted that for $n_{liq.} < n_w$ the correction in question is compensated by the change in brightness of the scattered light on emergence from the vessel. According to our data, however, the first correction is less than 1%, whereas the second may reach 50%. It is hard to explain such a sharp difference between our results and those of Bernstein and Allen. It is not impossible that this is related to inaccuracies in the theoretical calculation of the latter work, no allowance being made for multiple reflections of the exciting light on emerging from the vessel [44, 45], or to the not irreproachable experimental method of measuring the reflection losses, which was based on the assumption that the relation between the concentration of the substance in solution and the Raman intensity was linear.

TABLE 1. Effect of Reflection on the Intensity of the Scattered Light

Substance in the gap of the vessel	n	n/n_W	I, rel. units	Mean square error, %
Air	1	0.66	0.96	1.5
Alcohol	1.36	0.89	0.98	1.5
Toluene	1.52	1.00	1.00	—
Carbon disulfide	1.67	1.10	0.99	1.5

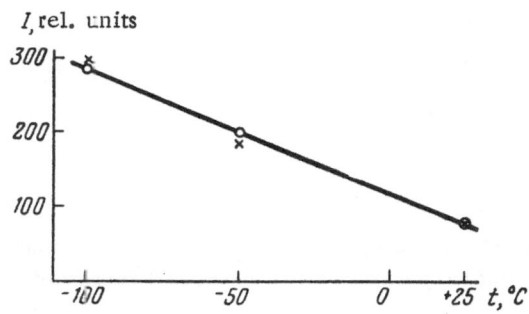

Fig. 2. Variation of the intensity of the 656 cm^{-1} line of carbon disulfide with temperature in a single lamp elliptical illuminator (crosses) and in an illuminator with diffuser (circles).

The effect of the configuration of the exciting rays on the measured value of the Raman intensity, and the effect of the conditions of imaging the vessel on the slit of the apparatus were checked for our system by two methods.

First Method. Measurements were made of the temperature dependence of the intensity of the 656 cm^{-1} line of carbon disulfide in a single lamp elliptic illuminator and in an illuminator with a diffuser, i. e., for radically differing configurations of the light falling on the vessel. As seen from Fig. 2, the results obtained in the two series of measurements are in complete harmony. Hence a change in the configuration of the exciting radiation and a change in the imaging of the vessel have practically no effect on the measured value of the scattered intensity.

Second Method. This was proposed by Rea [44] and consisted of graduating the experimental system with a standard one for which the relation between the intensity of the scattered flux and the refractive index is subject to theoretical calculation. As standard we selected a vessel in the form of a rectangular parallelepiped with a square exit window. The vessel was illuminated by an almost parallel beam of light. In this case the illumination in the sample does not depend on the refractive index of the liquid.

The experiments were carried out in a quartz vessel made from plane parallel plates. The aperture of the exciting radiation was limited both in the vertical and horizontal planes by means of diaphragms. In this vessel we compared successively the intensities of the lines of two substances with different refractive indices (carbon disulfide, n_{liq} = 1.6748, acetonitrile n_{liq} = 1.3441). As unit of comparison we took the intensity of the 2942 cm^{-1} line of acetonitrile. Then the intensities of the same lines were determined in the cylindrical vessel without aperture diaphragms, i. e., under working conditions. The results for the line intensities of carbon disulfide and acetonitrile are shown in Table 2, where I_1 and I_2 are the line intensities of carbon disulfide and acetonitrile respectively. As seen from the table, the relative intensities of the lines under the different experimental conditions agree with each other, not depending on the refractive index of the scattering liquid. Hence in our experiments a change in the configuration of the exciting rays or in the conditions of imaging the vessel at the slit of the spectrograph has no effect on the measured value of the intensity.

Thus in our apparatus we only took account of the change in the brightness of the scattered light on emergence from the vessel, proportional to n_{liq}^2. In particular, in measuring the intensity ratio of the etonitrile and carbon disulfide lines the correction was $(1.67/1.34)^2$ = 1.55. It is an important point that this correction is determined only by n_{liq} and not by n_W. The rest of the factor require the introduction of no corrections.

Thus on the basis of the foregoing considerations the intensity of Raman lines, referred to a gram molecule of the substance, is given by the formula

TABLE 2. Intensity Ratio of Carbon Disulfide and Acetonitrile Lines in Vessels with Square and Circular Cross Sections

Vessels	Substance	Δv, cm^{-1}	I_1/I_2	n_{liq}
Vessel with square cross section	Carbon disulfide	656 664	4.74 ± 0.08 1.54 ± 0.08	1.67
	Acetonitrile	2942	1.00	1.34
Vessel with circular cross section	Carbon disulfide	656 664	4.92 ± 0.2 1.50 ± 0.1	1.67
	Acetonitrile	2942	1.00	1.34

$$I = kI_0 \frac{Mn_{liq}^2}{d}, \qquad (9)$$

where I_0 is the observed intensity allowing for the spectral sensitivity of the apparatus, k is a constant depending on the choice of measuring units, and M, n, d are the molecular weight, refractive index, and density of the scattering substance.

We note that the correction n_{liq}^2/d is important when the substance passes from the liquid phase into the vapor, or on comparing line intensities of different substances. For temperature measurements, the change in n_{liq}^2 is compensated by the change in density of the substance, since the ratio n_{liq}^2/d is practically independent of temperature.

The study of such important parameters as the shape and width of the Raman lines also requires allowance to be made for the distorting effect of the apparatus on the quantities measured.

The most rapid and efficient way of obtaining the true shape of Raman lines from the shape observed consists of making separate and successive allowance for the effects of slit and nonslit distortions. The main advantage of the approach lies in that the apparatus function does not depend on the slit width, and is determined only once for a given apparatus. During measurements, the slit width may be selected to correspond to the parameters of the line being examined. If the slit width is less than $\frac{1}{3}$ of the line width, the slit corrections may be neglected. Thus in our papers [46, 47], finding the true contour and line width from those observed reduced to the following operations: 1) distortion introduced by the finite slit width of the apparatus was eliminated from the observed contour by Rayleigh's method; 2) the nonslit apparatus function, previously determined by recording the exciting line with a slit width less than normal, was eliminated from the Raman line shape with infinitely narrow apparatus slit. In our case, with photographic recording, the nonslit apparatus function was close to a dispersion curve of width 0.5 cm^{-1}. Since the observed contour of the majority of the Raman lines after eliminating the slit distortions were approximated by a dispersion curve, the true contour in this case was also a dispersion curve with a width equal to the difference between the widths of the observed contour and the apparatus function.

With the photoelectric method of recording, the observed contours of the Raman lines, as a rule, constitute a convolution of a dispersion curve (true contour of Raman line) and a Gauss curve (nonslit apparatus function). In this case the true Raman line width was determined from a graph giving the connection between the widths of the true contour and apparatus function and the width of their convolution [48].

§2. Apparatus for Recording the Contours of Raman Effect Lines

The problem of determining the true line parameters from those observed may be simplified if the experimental conditions are so chosen that the observed contour lies close to the true one. For this, measurements must be carried out in systems with large dispersion and resolving power. As the intensity of the Raman spectra is not great, the spectral systems used must have good transmission. We made our measurements in the diffraction spectrometer DFS-4 and French three-prism spectrograph HUET B-III, which satisfied these requirements. In the DFS-4 system, a plane diffraction grating with 1200 lines/mm and ruled area 100 by 90 mm² concentrated 52% of the reflected light into the first order. The linear dispersion of the spectrometer in the first order equalled 6.4 A/mm.

As receiver for the light signal we used a photomultiplier of the FÉU-17 type with an antimony—cesium photocathode. The photomultiplier with the best signal-to-noise ratio was selected experimentally. The rated data of the FÉU-17 were: integral sensitivity 7.5 A/lm, dark current $3 \cdot 10^{-11}$ A at voltage 800 V. The most favorable signal-to-noise ratio was obtained at 700 V.

The photoelectric part of the DFS-4 apparatus was somewhat modernized. The single-tube direct-current amplifier was replaced by an ÉMU-3 electrometric amplifier constituting a three-cascade direct-current amplifier with electrometric output incorporating 100% negative feedback.

After amplification the signal was recorded by the automatic ÉPP-09 recorder. For better averaging of noise in recording the wide, low-intensity Raman lines, an additional RC-network was attached to the input of the ÉPP-09, changing the time constant of the writing system from 5 to 40 sec. A number of the parameters of the receiver-amplifier system important for our work were determined experimentally. We measured the signal at the output of the system as a function of the load resistance. It was found that the amplification of the ÉMU-3 varied from $2.5 \cdot 10^5$ to $9.1 \cdot 10^8$.

For control of the linear relation between the signal at the input and output of the system, we obtained the relation between the signal being recorded and the monochromator slit width. The spectral sensitivity of the monochromator together with the photomultiplier was studied with the aid of the continuous fluorescence spectrum of a solution of quinine sulfate ($5 \cdot 10^{-5}$ g/ml) in water with a known spectral distribution of intensity [45].

Calibration of the spectral sensitivity of the system was performed with various entrance and exit slits of the monochromator. In this the dispersion curve of the apparatus was taken into account, since the geometric slit width remained constant during the measurements. Ths spectral characteristics of the apparatus so obtained are presented in Fig. 3.

The experimentally measured minimum time constant of the recording system was 5 sec. The grating rotation rate v could be varied from 0.25 to 4.2 A/min.

It follows from the condition for a minimum of the systematic and random errors in the photoelectric recording of the spectrum that the time t for the line width to pass over the exit slit of the monochromator

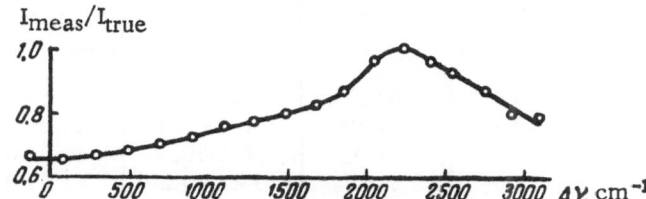

Fig. 3. Spectral sensitivity of spectrometer DSF-4 in the blue part of the spectrum. Abscissas give light frequency (blue line of mercury taken as zero); ordinates give the ratio of the intensity of the quinine luminescence spectrum measured on the DFS-4 to the true value.

must be considerably greater than the time constant τ of the recording system (t $\approx$ 10τ) [49]. In our system (τ = 5 sec, v = 0.25 A/min), lines of breadth reaching 0.8 cm^{-1} could be recorded under the conditions stated.

The observed width of the mercury 4358 A line recorded with entrance and exit slits of 0.07 cm^{-1} was 0.8 cm^{-1}. Since the true width of the 4358 A line for a low-pressure lamp was (according to [50]) 0.2 cm^{-1}, the width of the nonslit apparatus function of the spectrometer was approximately 0.6 cm^{-1}.

In studying the Raman spectra, the slit widths, recording rate, and the time constant of the recording system were selected in accordance with the observed line width. In studying narrow, intense lines like that of benzene at frequency 992 cm^{-1} we used slits of width 0.5 cm^{-1}, the time constant was 5 sec, and the slowest scanning rate was employed. For weak, wide lines, the slits varied from 3 to 5 cm^{-1}, the time constant from 5 to 40 sec, and the scanning rate from 0.25 to 4.2 A/min respectively.

The changes in the photoelectric part of the apparatus mentioned above, and suitalbe choice of the ratio between the rate of rotation of the diffraction grating and the motion of the paper in the automatic recorder, made it possible to obtain reliable measurements of the Raman line parameters in which we were interested (the shape, integral intensity, and width of the lines).

Together with the DFS-4 diffraction spectrometer with photoelectric recording, we also used extensively the French three-prism spectrograph HUET B-III with photographic recording. The relative aperture of the spectrograph was 1:6. The dispersion in the 4358 A region was 8 A/mm, and in the 4047 A region 6 A/mm. The casing of the spectrograph was furnished with double walls, between which circulated thermostating liquid. For recording the Raman spectra we used the high-sensitivity, fine-grained plates "Raman platten" and "Kodak 0-ao." The blackening on the photoplate was converted into intensities with the help of a graduated reducer, the transmission of the steps in this being measured beforehand. The reducer was calibrated under working conditions. The spectrograms obtained were measured in the MF-2 microphotometer with magnification 27.

Since two spectral systems were employed in our investigations, we were able to check how well the results obtained in different systems were reproduced, and at the same time to estimate the measuring errors.

Table 3 shows results of measuring the integral intensities and widths of a number of cyclohexane lines obtained in the diffraction spectrometer DFS-4 and the spectrograph B-III. The experimental data were obtained with a scattering liquid temperature of 25°C by averaging measurements from eight spectrograms.

As we see from Table 3, the data obtained in different systems and treated in the way described above (see Chapter II, §1, 2) agree closely with each other. The mean arithmetical error in determining the integral intensity of the most typical lines presented was around 5% by the photoelectric method of recording and 8% by the photographic method; the errors in determining breadths by these two methods were around 3% and 5% respectively.

TABLE 3. Parameters of the Raman Lines of Cyclohexane Obtained in the DFS-4 and B-III

Δv, cm^{-1}	I_∞, rel. units		δ, cm^{-1}	
	DFS-4	B-III	DFS-4	B-III
802	91±5	93±7	1.9±0.1	2.1±0.2
1029	96±4	94±8	10.5±0.3	10.8±0.5
1267	100	100	10.8±0.2	11.0±0.5
1445	93±5	96±7	13.2±0.2	13.5±0.7

CHAPTER III

EXPERIMENTAL STUDY
OF THE TEMPERATURE DEPENDENCE
OF RAMAN LINE INTENSITIES
FOR SUBSTANCES IN THE LIQUID STATE

§1. Method of Investigation

The objects of investigation chosen for the present work were substances belonging to various classes of compounds and not entering into any specific intermolecular interactions in the liquid state (hydrogen bonds, complexing, electrochemical processes, etc.). All the substances were previously purified by repeated vacuum distillation.

In order to obtain Raman spectra at various temperatures we constructed a special container. A section of this is shown in Fig. 4. The vessel 1 containing the scattering liquid was surrounded by two glass sleeves 2 and 3. Between the vessel and the first sleeve 2 passed hot air for work in the high-temperature range, or cold nitrogen obtained by evaporating liquid nitrogen for work in the low-temperature range. The vacuum between the first and second (outer) sleeves 2 and 3 protected the walls and the plane parallel windows 4 of the cuvette from condensation.

The use of this vessel enabled us to study the temperature dependence of the Raman spectra of substances in the range −140 to +150°C. The temperature of the substance under examination was measured by means of a copper−constantan thermocouple introduced into the working part of the cuvette. Special investigation showed that the fall in temperature along the working part of the cuvette was no more than 1 to 2°C.

The cuvette containing the substance to be investigated was placed in a two-lamp illuminator. As light source we used low pressure mercury lamps. The lamp supply came from a constant-voltage network (110 V) with a working lamp current of 10 A. As exciting radiation we used the lines Hg 4047 A and Hg 4358 A. For separating out the exciting lines in a number of cases we used glass filters. In order to control the reproducibility of the experimental results, the measurements were carried out both on lowering the temperature and on raising it. The spectrograms were treated by the method described in Chapter 2, §1. The intensity of a line at a given temperature was expressed in terms on the intensity of the same line at 25°C, taken as 100. In cases in which the relative intensities inside the spectrum were measured at various temperatures, the intensity of the strongest line in the spectrum at 25°C was taken as 100.

§2. Temperature Dependence of the Intensity of the Stokes Lines in the Raman Effect

It was established from a large amount of experimental material that, in every one of the liquids which we investigated, without exception, the intensity of the Stokes lines in the Raman effect fell with increasing temperature. This result agrees qualitatively with the results of the majority of earlier investigations [12, 13, 18-29]. The relation between the line intensity and temperature is nearly linear.

TABLE 4. Relative Intensities of Raman Lines of Benzene at Various Temperatures

$\Delta\nu$, cm^{-1}	I, rel. units		$\Delta\nu$, cm^{-1}	I, rel. units	
	25° C	70° C		25° C	70° C
607	15	13	1606	10	9
850	6,3	5	2951	6	5
992	100	74	3047	54	48
1178	18	15	3062	80	70
1586	16	14			

The observed fall in intensity is evidently not linked with any redistribution of energy within the spectrum of the substance. This was checked experimentally by a simultaneous study of the temperature dependence of the intensity of all the main lines in the spectra of a number of substances. By way of example, Fig. 4 shows results of measuring all the main lines of benzene at temperatures 25 and 70°C. As seen from the table, the intensity of all lines in the Raman spectrum of benzene falls with rising temperature.

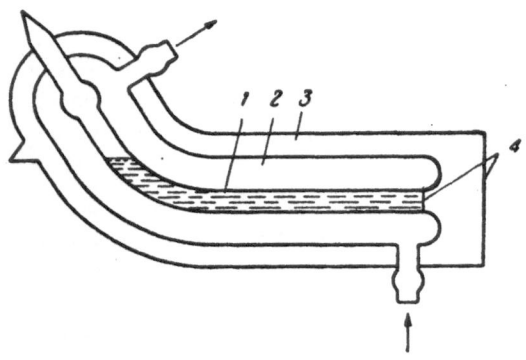

Fig. 4. Vessel for studying Raman spectra of liquids at various temperatures. 1) Cuvette containing scattering liquid; 2) sleeve for heating or cooling the substance; 3) Dewar; 4) plane parallel windows.

Our investigations showed that the form of the vibrations and the degree of depolarization of the lines ρ did not determine the extent of the temperature variations of Raman line intensities. By way of example, Table 5 shows results of measuring the Raman line intensities of benzene, toluene, and hexylbenzene.

According to calculations by A. M. Bogomolov [51], deformation-valence ring vibrations 607, 1586, and 1606 cm^{-1} are characteristic in form and the displacement of the atoms for benzene and its monosubstitution products. If the observed effect is determined only by the form of the molecular vibrations, the intensity changes of the corresponding line must be approximately the same for all the liquids considered. In the benzene and toluene spectra, the changes in the lines indicated with temperature are approximately the same, reaching some 30% at 100°C. In the hexylbenzene spectrum, however, these lines hardly change at all with temperature. In exactly the same way, the changes in the intensity of the valence-deformation ring vibrations 1004 and 1031 cm^{-1} of the toluene molecule are greater than the intensity changes of the corresponding vibrations in hexylbenzene.

In Table 6 we compare the temperature dependence of intensity for lines of substances similar in structure and containing a double bond $C = C$. We see from the table that the change in intensity of the valence vibration of the double bond 1642 cm^{-1} of heptene-1 and diallyl, and other lines relating to different forms of vibrations, are approximately the same. In molecules with other structures (tetramethylethylene and 3-ethylpentene-2), the changes in the intensity of the $C = C$ bond valence vibrations are less than in the case of heptene-1 and diallyl.

From the example given in Tables 5 and 6 we may be convinced that the temperature changes in the line intensities do not depend on their degree of depolarization. Thus, from Table 5 we see that, in the benzene spectrum, the intensity of the polarized line 992 cm^{-1} ($\rho = 0.11$) changes more with temperature than that of the depolarized lines 607 cm^{-1} ($\rho = 0.88$), 1586 cm^{-1} ($\rho = 0.81$), and 1606 cm^{-1} ($\rho = 0.80$). In the

TABLE 5. Intensity of Stokes Lines in the Raman Effect of Benzene and Its Monosubstitution Products at Various Temperatures

Substance	Δv, cm^{-1}	ρ	I, rel. units				$\Delta I,\%$ at 100° C
			25° C		70° C		
Benzene	607	0.88	15		13		28
	992	0.11	100		74		52
	1586	0.81	16		14		26
	1606	0.80	10		9		26
			—10° C	25° C	60° C	100° C	
Toluene	521	0.61	33	30	29	27	18
C	623	0.73	25	23	23	19	27
	786	0.09	63	57	53	50	19
	1004	0.07	110	100	94	89	17
	1031	0.10	32	29	27	26	16
	1211	0.13	35	35	33	31	15
	1586	0.76	19	17	16	14	25
	1606	0.70	42	38	35	31	25
			25° C	80° C	125° C		
Hexylbenzene	623	0.90	21	20	19		7
C—C—C—C—C—C	1003	0.11	100	96	94		6
	1031	0.09	34	34	33		4
	1584	0.77	19	18	18		6
	1606	0.76	57	55	53		7

toluene spectrum, the changes in intensity of the polarized lines (Δv = 786 cm^{-1}, ρ = 0.09; Δv = 1004 cm^{-1}, ρ = 0.07; Δv = 1031 cm^{-1}, ρ = 0.10; Δv = 1211 cm^{-1}, ρ = 0.13) are approximately the same and equal to the changes in intensity of the depolarized line 521 cm^{-1} (ρ = 0.61).

In the present investigation, we measured the Raman line intensities at various temperatures for a number of substances differing in the magnitude of the dipole moment μ. The results of these measurements are given in Table 7. We see from this that the intensity changes with temperature most for carbon disulfide molecules (μ = 0). The fall in intensity for the spectra of carbon tetrachloride (μ = 0) and chloroform (μ=1.15), and for acetone (μ = 2.73) and acetonitrile (μ = 3.94) are roughly the same. Hence the dipole moment of the scattering molecules does not determine the magnitude of the observed effect.

Comparison of the experimental results obtained for molecules of different structure shows that on the average, as a rule, the temperature dependence of the intensity of the spectrum diminishes as the molecular structure becomes more complex. Thus the intensity change for the benzene spectrum at 100°C is approximately 30%, while for the more complex hexylbenzene molecule (see Table 5) it is 6%. The comparison between the temperature dependence of the spectral intensities of cyclopentane and 1,2-dicyclopentylethane, cyclohexane and 1,2-dicyclohexylethane, and dipropenyl and alloocimene presented in Table 8 confirm this law. We see from the table that the spectral intensity of cyclopentane changes by approximately 40% on raising the temperature to 100°C, while that of dicyclopentylethane, a more complex molecule belonging to the same homologous series, changes by no more than 20%. The spectral intensity of cyclohexane falls by some 20% over same temperature range, while that of 1,2-dicyclohexylethane remains practically constant. The spectral intensity of dipropenyl changes more with temperature than does that of alloocimene. It should be noted that the law under consideration is quite general, and is fulfilled in the majority of cases, even when the molecules being compared do not belong to the same class of compounds. We shall deal with this point in more detail in subsequent sections.

TABLE 6. Temperature Dependence of Raman Line Intensities for Certain Unsaturated Hydrocarbons

Substance	Δv, cm^{-1}	ρ	I, rel. units				ΔI, % at 100° C
			−10° C	25° C	50° C	90° C	
Heptene-1 C=C—C—C—C—C	1294	depol.	111	98	90	79	28
	1304	depol.	111	98	90	79	28
	1439	depol.	95	84	80	67	30
	1642	0.15	115	100	91	77	36
			−70° C	−20° C	25° C	50° C	
Diallyl C=C—C—C—C=C	1299	depol.	118	104	88	84	32
	1416	depol.	88	75	64	59	29
	1642	0.12	138	115	100	88	40
			−50° C	−10° C	25° C	70° C	
Tetramethylethylene	504	0.70	40	34	34	30	20
	690	0.15	100	105	100	85	15
	1683	0.13	111	103	95	85	22
			−10 °C	25° C	60° C	100° C	
3-ethylpentene-2 C—C=C—C—C	724	0.10	30	27	26	24	19
	1440	0.74	112	100	94	84	27
	1459	0.74	112	100	94	84	27
	1670	0.20	90	83	85	79	11

§3. Experimental Results of Measuring the Stokes and Anti-Stokes Spectral Regions and Comparison with Theory [52, 53]

According to the statistical theory of the temperature dependence of Raman spectral intensities, the intensities of the Stokes and anti-Stokes lines should rise on increasing the temperature; their ratio in regions well removed from resonance should obey formula (3). However, as we saw in §2 of this Chapter, the intensity of the Stokes lines in the Raman effect decreases with increasing temperature, thus contradicting elementary theory. The validity of elementary theory for the liquid phase may be ascertained by studying the ratios of the Stokes and anti-Stokes satellite intensities at various temperatures. As we know, these ratios directly characterize the relative populations of the zero and first vibrational levels at the given temperature. Since, according to elementary theory, the temperature changes in the intensities of the lines depend on the magnitude of the vibrational quanta, as objects of investigation we selected substances containing fairly intense lines over a wide range of frequencies.

We studied the temperature dependence of Raman line intensities both in the Stokes and anti-Stokes regions of the spectrum for the following liquids: tribrommethylgermanium, * carbon tetrachloride, chloroform, carbon disulfide, and benzene. The experimental results obtained are shown in Table 9. We see from these that the intensity of the Stokes satellites falls with increasing temperature for all the substances, contrary to elementary theory [see formula (1)]. Thus, for example, the intensities of the GeCH$_3$Br$_3$ lines at 96

---

* The GeCH$_3$Br$_3$ was synthesized and kindly presented to us for study by G. Ya. Zueva.

TABLE 7. Temperature Dependence of Spectral Intensities for Molecules with Different Dipole Moments

Substance	Δv, cm^{-1}	μ, Debye	I, rel. units				ΔI, % at 100°C
Carbon disulfide			—100 C°	—50 °C	25 °C		
	656	0	288	198	100		160
	796		52	39,5	20		140
Carbon tetra-chloride			—20 °C	25 °C	70 °C		
	217	0	100	87	75		28
	314		105	97	83		25
	459		104	100	85		22
Chloroform			—40° C	25 °C	60 °C		
	366	1.15	114	98	88		26
	665		110	100	94		16
Acetone.			—70° C	—40° C	25 °C	50 °C	
	786	2.73	115	104	87	78	30
	1642		143	123	100	86	46
	1707		89	77,5	62	52	46
Acetonitrile			—40° C	25 °C	70 °C		
	384	3.94	36	29	24		40
	918		23	20	17		30
	2256		85	70	59		38
	2942		114	100	87		27

and 128 cm^{-1} fall by 30% on raising the temperature from 25 to 150°C, whereas according to theory they should rise by 37 and 32% respectively. The intensities of the 264 and 315 cm^{-1} lines fall by 23 and 40% for the same temperature change, instead of rising by 22 and 20%.

A rather different picture holds for the anti-Stokes lines. The intensity of the anti-Stokes lines either remains unchanged or rises with temperature, but always more slowly than theory would deman [see formula (2)]. As an example, let us consider the GeCH$_3$Br$_3$ lines given in Table 9. The intensities of the anti-Stokes lines 96 and 128 cm^{-1} are practically unvarying with temperature, whereas according to theory they should rise by approximately 50% on increasing the temperature from 25 to 150°C. The intensity of anti-Stokes line 264 cm^{-1} rises by 30% instead of 77%. Analogous behavior is found for the line intensities of the other substances shown in the table.

In order to ascertain the validity of the statistical treatment of the temperature variation of Raman line intensities for liquids, it is essential, as indicated above, to consider the ratio of the intensities of the Stokes and anti-Stokes components. Study of the spectra of carbon tetrachloride, chloroform, benzene, and tribrommethylgermanium over a wide temperature range showed that this ratio was close to that calculated from formula (3).

It should be noted that the self absorption of the substance exerts considerable influence on the observed Stokes-to-anti-Stokes intensity ratio in the case of carbon disulfide. A correction for absorption is given by a formula derived in papers by Michel, Behringer [54, 55], and others. The true Stokes-to-anti-Stokes intensity ratio will in this case be

TABLE 8. Changes in Spectral Intensity with Temperature as a Function of the Struc-
of the Molecule

Substance	$\Delta\nu$, cm⁻¹	ρ	I, rel. units				ΔI, % at 100° C
			−80° C	−40° C	25° C	40° C	
Cyclopentane	889	0.12	133	120	100	95	32
	1031	0.82	74	66	54	51	37
	1449	0.72	80	70	60	56	50
			25° C	80° C	120° C		
1,2-dicyclopenthyl-ethane	894	0.12	64	57	52		19
	1450	0.87	100	84	80		21
			25° C		80° C		
Cyclohexane	798	pol.	91		83		16
			96		81		23
	1032	depol.	100		85		22
	1443	depol.	93		80		25
			25° C	60° C	125° C		
1,2-dicyclohexyl-ethane	802	0.13					
	1029	0.79	34	34	83		2
	1267	0.77	80	79	80		0
	1445	0.81	100	101	97		3
			−50° C	25° C	80° C		
Dipropenyl	434	0.68					
C—C=C—C=C—C	1378	0.47	7.5	6	5.1		36
	1450	0.58	13.3	11	8.7		31
	1657	0.35	18	14	12		29
	1668	0.35	130	100	80		40
			74	57	45		40
			25 °C	100° C			
Alloocimene[b]	1151	depol.					
C—C=C—C=C—C=C—C	1184	depol.	10.3	9.0			15
C C	1235	depol.	8.2	7.5			10
	1593	pol.	7.2	6.2			18
	1629	pol.					
	1643	pol.	100	87			17

[a] At temperature −80°C the 889 cm⁻¹ line of cyclopentane splits into two components. The splitting may be connected with the presence of isomeric forms of cyclopentane in the liquid.

[b] It is known that dipropenyl and alloocimene in the liquid state contain a set of rotational isomers. In the temperature range studied, no redistribution of intensity was observed in the spectrum of the substances.

TABLE 9. Temperature Dependence of the Raman Line Intensities in the Stokes and Anti-Stokes Parts of the Spectrum (I in rel. units)

$\Delta\nu$,cm	t, °C	I_{St} Experiment	I_{St} Theory	I_{ASt} Experiment	I_{ASt} Theory	$\dfrac{I_{St}}{I_{ASt}}$ Experiment	$\left(\dfrac{\nu-\nu_j}{\nu+\nu_j}\right)^4 e^{\frac{h\nu_j}{kT}}$
				GeCH$_3$Br$_3$			
96	+25	79	79	47	52	1.7	1.5
	+150	56	108	43	81	1.3	1.3
128	+25	25	25	14	14	1.8	1.8
	+150	18	33	13	22	1.4	1.5
164	+25	30	30	17	15	1.8	2.0
	+150	25	38	14	23	1.8	1.6
264	−70	—	—	—	—	5.0	6.5
	−25	—	—	—	—	3.8	4.2
	+25	100	100	34	30	2.9	3.3
	+150	77	122	44	53	1.8	2.3
315	+25	25	25	—	—	—	—
	+150	15,5	30	—	—	—	—
				CCl$_4$			
217	−20	100	81	31	25	3.2	3.2
	+25	87	87	33	32	2.6	2.7
	+70	75	96	37	40	2.0	2.4
314	−20	105	91	19	17	5.5	5.4
	+25	97	97	23	24	4.2	4.0
	+70	83	104	26	31	3.2	3.4
459	−20	104	96	9	8.3	11.6	11.5
	+25	100	100	11	11	9.1	9.1
	+70	85	104	12	18	6.8	5.8
				CHCl$_3$			
366	−40	114	91	13.7	10.6	8.3	8.6
	+25	98	98	20	19	4.8	5.1
	+60	88	103	22	24	4.0	4.3
665	+25	100	110	5.1	4.9	19.4	20.4
	+60	94	103	7	7.4	13.5	14.0
				C$_6$H$_6$			
992	+25	100	100	1.1	1.15	91	87
	+70	74	103	1.8	2.3	42	45
				CS$_2$			
656	−50	198	98	2.4	1.8	80	66
	−20	155	99	3.4	3.0	46	39
	+25	100	100	7.1	5.2	14	12

$$\frac{I_{St}}{I_{ASt}} = \frac{I_{St.o} k_{St} (1 - e^{-k_{St} l})}{I_{ASt.o} k_{ASt} (1 - e^{-k_{St} l})}, \tag{10}$$

where $I_{St.o}$ and $I_{ASt.o}$ are the observed intensities of the Stokes and anti-Stokes lines respectively, k_{St} and k_{ASt} are the absorption coefficients at the wavelengths of the Stokes and anti-Stokes lines,* and l is the length of the illuminated part of the scattering sample.

Thus the experimental data presented indicate that the statistical theory of the temperature variation of Raman intensities is valid in the liquid phase.

The temperature dependence of the Stokes and anti-Stokes line intensities in the Raman effect shows that, for the liquid, together with the temperature redistribution of molecules with respect to the vibrational levels, there is another factor which leads to a general weakening of the spectral intensities as the temperature rises. In view of the fact that the intensity of the Stokes components, even for low-vibration frequencies, depends according to theory only weakly on temperature, this factor leads to a change in the sign of the temperature vibration of the Stokes components. For the anti-Stokes components, the intensity of which is most sensitive to temperature change, there is also a fall in intensity, although experiment remains in qualitative agreement with theory.

On the basis of what has been said, we may assume that the most probable cause of the anomalous temperature dependence of Raman intensities in liquids lies in the effects of intermolecular interactions.

---

*For measuring the absorption coefficient we used the system and method developed by M. M. Sushchinskii and V. A Zubov. We acknowledge the helpful cooperation of these workers.

CHAPTER IV

EXPERIMENTAL STUDY
OF THE TEMPERATURE DEPENDENCE
OF RAMAN LINE INTENSITIES
FOR SUBSTANCES IN THE VAPOR STATE [56, 57]

In order to ascertain whether the phenomenon of the anomalous temperature dependence of line intensities is connected with molecular interactions, we need information on vapors at low pressures. Unfortunately, in view of the great methodical difficulties involved, there are no such data in the literature. Only recently has there appeared a paper by Ya. S. Bobovich and V. M. Pivovarova [58] in which the line intensities of gaseous nitrogen and carbon dioxide were measured at various temperatures. However, the small temperature effect, as well as the absence of data on the temperature dependence of line intensities for nitrogen and carbon dioxide in liquid form, does not allow us to draw unequivocal conclusions regarding the role of intermolecular interactions in the phenomenon observed. In view of this, we regarded it as appropriate to study the temperature dependence of Raman spectral intensities for one and the same substance in both gaseous and liquid states. For investigation we selected substances having fairly intense lines over a wide frequency range.

§1. Methods of Investigation

As subjects for investigation we chose carbon tetrachloride, chloroform, acetone, acetonitrile, benzene, carbon disulfide. The substances were carefully purified and sealed into a cuvette under vacuum. The cuvette was a pyrex tube 650 mm long with a 30 mm diameter plane parallel window. We used simple construction for the cuvette, since measurements of the spectral intensity of vapor in a multiple-pass cuvette are accompanied by systematic errors connected, in particular, with defocusing and a change in the reflection coefficient of the mirrors. The temperature changes were carried out at constant vapor pressure in the cuvette. For this purpose the cuvette was furnished with a side tube containing the liquid under investigation at constant temperature. In our experiments the vapor pressure was 2 atm. The working part of the cuvette was heated.

The temperature of the illuminated part of the cuvette was maintained by means of a nichrome spiral wound on a pyrex glass tube. In order to reduce the fall of temperature along the axis and cross the diameter of the furnace, the density of turns at the ends of the furnace was greater than in the middle, and the cuvette was displaced slightly downwards relative to the furnace axis. The fall in temperature along the axis of the cuvette and over the height was measured by means of a differential thermocouple. The maximum temperature drop over the height was 3 to 4°C, and along the length 5 to 6°C. The temperature of the working part of the cuvette was also measured by a thermocouple. The side tube of the cuvette was heated in a beaker full of glycerine. The temperature of the glycerine was measured by a thermometer. In a second variant (Fig. 5), the furnace consisted of a Dewar vessel with plane parallel windows, through which hot air was blown.

The cuvette containing the substance for examination was placed in a two-lamp illuminator, the inner walls of which were covered by a 2 to 3 mm layer of magnesium oxide. The vessel was so adjusted that the light reflected from its walls should not fall into the spectrograph slit (see Chapter 2, §1). As light sources

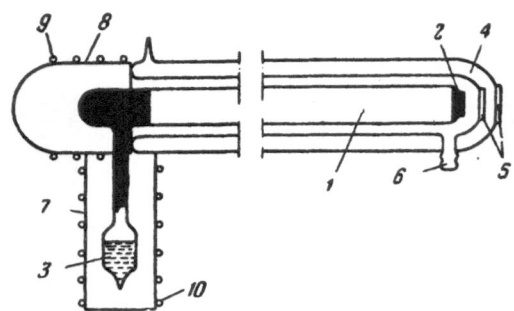

Fig. 5. Cuvette for studying Raman spectra of vapors. 1) Working part of the vessel; 2) plane parallel window; 3) cuvette side tube containing the liquid of the substance being studied; 4) Dewar; 5) plane parallel window of the Dewar; 6) hot air supply tube; 7) quartz beaker containing glycerine; 8) pyrex furnace; 9, 10) heating coils.

we used low-pressure meter lamps of the Stoicheff type [59]. The working conditions of the lamp were: voltage 110 V, current 20 A. The discharge tube of the lamp was continuously cooled by a stream of water. In order to avoid condensation of mercury in the working part of the channel, the temperature of the water in the central sleeve (Fig. 6) was 20°C higher than that of the water at the ends of the lamp. Under working conditions, the breadth of the exciting line 4358 A, together with the apparatus function, equalled 0.8 cm^{-1}.

Measurements were made both photoelectrically and photographically. The recording of the spectra began some hour and a half after the desired temperature had been established. For control of the reproducibility of the experimental results, measurements were made both when the temperature was being raised and when it was being lowered. The results accepted were the mean of those given by four to six spectrograms. The integral intensity at various temperatures was determined by planimetering the line contours over a frequency range equal to the same number of half-widths. The true intensity belonging to a molecules of the substance was found from the formula

$$I = I_0 \, kTn^2/p, \tag{11}$$

where I_0 is the observed intensity, k Boltzmann's constant, T the absolute temperature, p the pressure of the vapor, n the refractive index. In our relative measurements, the intensity of the line at the lowest temperature was taken as 100.

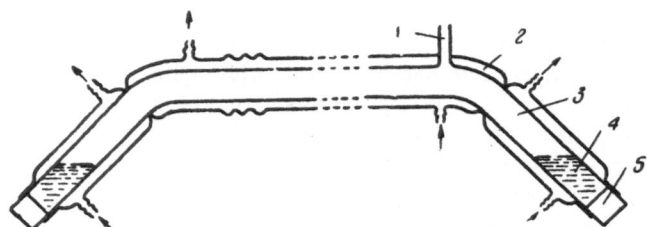

Fig. 6. Low-pressure mercury lamp of the Stoicheff type. 1) Side tube for sealing lamp to vacuum system; 2) glass sleeve with circulating water for cooling the discharge; 3) mercury glow discharge; 4) mercury; 5) kovar electrode.

§ 2. Temperature Dependence of Raman Line Intensities in Vapors

The results of measuring the temperature dependence of the line intensities of carbon tetrachloride, chloroform, acetone, acetonitrile, benzene, and carbon disulfide in the vapor phase are shown in Tables 10 to 12. The experimental data are compared with theory. The same tables give data on the temperature dependence of the spectral intensities of the liquids. Table 10 shows only the experimental results for the Stokes lines. We see from this that the Stokes lines in the spectra of all the substances studied in the vapor state increase in intensity or remain constant as the temperature rises. In the liquids, the intensities of these lines fall on increasing the temperature. Thus, for example, in gaseous carbon disulfide the intensity of the line $\Delta\nu$ = 656 cm^{-1} remains constant to within experimental error on changing the temperature by 44°, whereas in the

TABLE 10. Temperature Dependence of the Intensities of the Stokes Lines for a Number of Substances in Liquid and Vapor Form (I in rel. units)

Substance	$\Delta\nu$,cm^{-1}	I_v			I_{liq}	
		Experiment		Theory	Experiment	
		78° C	122° C	122 °C	0° C	40° C
Carbon disulfide	656	100	108	102	100	52
		120 °C	200 °C	200 °C	—40 °C	60° C
Acetone	786	100	106	104	100	70
	2922		109	100		75
		110° C	190° C	190° C	25° C	70° C
Benzene	992	100	110	103	100	74
	3062		104	100		80
		120° C	200° C	200° C	—30° C	70° C
Acetonitrile	2942	100	98	100	100	83

TABLE 11. Temperature Dependence of Raman Line Intensities in Liquid and Gaseous Carbon Tetrachloride (I in rel. units)

$\Delta\nu$, cm^{-1}	t, °C	I_{St}		I_{ASt}		I_{St}/I_{ASt}	
		Experiment	Theory	Experiment	Theory	Experiment	Theory
Vapor							
314	112	87	87	26	25	3.3	3.2
	190	103	97	38	40	2.7	2.4
459	112	100	100	20	21	4.9	4.8
	190	117	108	30	30	3.9	3.6
Liquid							
314	—20	100	100	18	19	5.5	5.3
	70	79	114	23	34	3.2	3.3
459	—20	100	100	9	9	11.6	11.6
	70	82	103	14	19	6.8	5.9

liquid it falls by 50%. For benzene in the gaseous state the intensity of the 992 cm^{-1} line rises approximately 10% on changing the temperature from 110 to 190°C. In liquid benzene the intensity of the same line falls 26% on changing the temperature from 25 to 70°C.

We see from Table 10 that the intensity change for the Stokes lines in vapor agrees fairly well with elementary theory. In order to convince ourselves that the statistical theory of the temperature dependence of Raman intensities is valid for the gaseous state, however, we must also know the changes of the anti-Stokes line intensities with temperature, and the ratio of the intensities of the Stokes satellites to those of the anti-Stokes satellites. Measurements of this kind in vapor are beset by considerable experimental difficulties. The problem is somewhat lightened by the fact that at high temperatures the intensities of the Stokes and anti-Stokes satellites tend to converge. We were able to measure the intensities of a number of lines in both Stokes

TABLE 12. Temperature Dependence of Raman Line Intensities in Liquid and Gaseous Chloroform (I in rel. units)

Δv, cm^{-1}	t, °C	I_{St}		I_{ASt}		I_{St}/I_{Ast}	
		Experiment	Theory	Experiment	Theory	Experiment	Theory
Vapor							
366	120	88	88	34	27	2.6	3.3
	200	133	97	58	46	2.1	2.6
665	120	100	100	—	—	—	—
	200	101	105	—	—	—	—
3018	120	82	82	—	—	—	—
	200	90	92	—	—	—	—
Liquid							
366	—40	98	98	12	11	8.3	8.6
	60	75	111	19	26	4.0	4.3
655	—40	100	100	—	—	—	—
	60	86	104	—	—	—	—
3018	—40	50	50	—	—	—	—
	60	39	50	—	—	—	—

and anti-Stokes spectral regions at various temperatures for carbon tetrachloride and chloroform vapors. The experimental results are given in Tables 11 and 12 respectively.

It follows from Table 11 that in gaseous carbon tetrachloride the intensities of the Stokes lines 314 and 459 cm^{-1} and also the anti-Stokes satellites rise with temperature in accordance with elementary theory. The results given in the table show that the intensity of the Stokes lines in the liquid phase falls with increasing temperature, while that of the anti-Stokes lines rises. The intensity ratios of the Stokes and anti-Stokes satellites given in the last two columns of the table coincide with theory both for the liquid and for the vapor phase.

As we see from Table 12, analogous laws are found in the spectra of liquid and gaseous chloroform. The intensity of the Stokes lines 366, 665, and 3018 cm^{-1} in the vapor rises and in the liquid falls on increasing the temperature. The intensity of the anti-Stokes satellite of line 366 cm^{-1} rises with temperature independently of the state of aggregation of the substance. The ratio I_{St}/I_{ASt} for the line 366 cm^{-1} in the liquid and gaseous phases coincides with that derived from formula (3).

The experimental results given above indicate that, in the case of weak intermolecular interactions such as in low-pressure vapors, the temperature dependence of the Raman spectral intensities does not contradict the elementary theory. These data lead us to the conclusion that, in the liquid phase, intermolecular interactions as well as the redistribution of the molecules with respect to the vibrational levels have an effect on the line intensities.

§3. Study of Raman Line Intensities on Transformation of the Substance from the Liquid to the Vapor State

In order to obtain as complete information as possible on the effect of intermolecular actions on the Raman intensities, we studied the changes in the intensity when the substance passed from the liquid phase into the vapor. It should be noted that at the present time there are no papers published on the effect of the state of aggregation of the substance on the Raman line intensities.

TABLE 13. Relative Intensities of Lines on Transformation from the Vapor to Liquid

Substance	$\Delta \nu$, cm^{-1}	t, °C	I_{exp}, rel. units	Substance	$\Delta \nu$, cm^{-1}	t, °C	I_{exp}, rel. units
	Vapor				Liquid		
CHCl$_3$ {	366	120	100	CHCl$_3$ {	366	60	140
	665	120	100		665	60	140
CCl$_4$	459	120	100	CCl$_4$	459	60	210
C$_6$H$_6$	992	120	100	C$_6$H$_6$	992	60	240
CS$_2$	656	80	100	CS$_2$	656	40	700

As subjects for investigation we selected carbon disulfide, benzene, carbon tetrachloride, and chloroform. We shall not stop to describe the experimental system and measuring procedure, as these were given quite fully earlier. We would add merely that in comparing the liquid and vapor spectral intensities we used identical cuvettes. The identity of the cuvettes was established by calibrating them with the help of Raman spectra of liquid benzene. The measurements were made in the diffraction spectrometer DFS-4 and the spectrograph B-III. In the photoelectric recording, the wide range of amplification of our electrometric amplifier enabled us directly to compare the intensities of the same lines for liquid and vapor. The amplification of the EMU-3 was first carefully calibrated. In the photographic recording method, in order to compensate the great difference in the intensities of the liquid and vapor spectra associated with the difference in density in these states of aggregation, we used a three step reducer. The transmission of the reducer was previously measured under working conditions.

Furthermore, for example, the intensity of the benzene 992 cm^{-1} lines in vapor was compared with that of the Stokes line 607 cm^{-1} and the anti-Stokes line 992 cm^{-1} in the liquid. Since the intensity of the lines in the liquid-benzene spectrum had been measured previously, it was not hard to find the ratio of the intensity of the 992 cm^{-1} Stokes line in vapor and liquid. The line intensities for the two phases of other substances were compared in the same way. The integral intensity of the line (per molecule) in liquid, measured relative to the integral intensity of the same line in vapor (taken as 100) equalled:

$$I_{liq} = \frac{I'_{liq}}{I'_{v}} I_{v} \ \frac{M n^2_{liq}}{d_{liq}} \frac{p}{kT} , \tag{12}$$

where I_v is the intensity of vapor at the given temperature, taken as 100, I'_{liq} and I'_v are the observed intensities of liquid and vapor, M is the molecular weight, n_{liq} and d_{liq} the refractive index and density of the liquid at the given temperature, k Boltzmann's constant, p the pressure of the vapor, and T the absolute temperature.

The mean scatter of the measurements of relative integral intensities was approximately 7 to 8%. The intensities of the Raman lines in the vapor of chloroform, carbon tetrachloride, and benzene were measured at 120°C, and that of carbon disulfide at 80°C; for the liquids, the corresponding temperatures were 60 and 40°C. The results of the measurements are shown in Table 13. As we see from the table, for all the substances investigated the Raman line intensities for the liquids as a rule exceeded the intensities for the vapors at given temperatures. Especially large were the intensity changes of the 656 cm^{-1} lines in the carbon disulfide spectrum. The intensity rose some 600% as the carbon disulfide passed from the vapor to the liquid state.

It would be more correct, however, to compare the line intensities in different phases at the same temperatures. The spectra of complex molecules may evidently be compared at different temperatures, since, as we showed experimentally, their line intensities change little with temperature. If we fail to allow for the temperature dependence of line intensities in the liquid in the case of simple molecules like carbon disulfide and benzene, we may be led to false conclusions. In order to obtain data for the same temperatures

TABLE 14. Relative Intensities of Lines on Passing from Vapor to Liquid at the Same Temperatures

Substance	n at $20°$ C	$\Delta\nu$, cm^{-1}	t, °C	I_{exp}, rel. units	I_{theor}, rel. units
Vapor					
CHCl$_3$	1.00	366	60	100	100
CCl$_4$	1.00	665	60	100	100
C$_6$H$_6$	1.00	459	120	100	100
CS$_2$	1.00	992	120	100	100
	1.00	656	40	100	100
			80		
Liquid					
CHCl$_3$	1.46	366	60	165	128
			120	120	125
		665	60	142	128
			120	124	125
CCl$_4$	1.47	459	60	220	129
			120	160	126
C$_6$H$_6$	1.52	992	60	250	133
			120	140	129
CS$_2$	1.70	656	40	740	154
			80	330	150

and different phases, we extrapolated the experimental curves relating line intensity and temperature for liquid and vapor. The relative changes in line intensity for the two phases of chloroform, benzene, carbon disulfide, and carbon tetrachloride are given as a function of temperature in Chapter 2, §2 and 3. As shown earlier (see Chapter 4, §2), the line intensities of the vapor vary little with temperature. According to our measurements, the spectral intensity of a number of liquids on heating above the boiling point falls in accordance with the laws established for all liquids. In view of this it is evident that extrapolation over 60°C will not introduce serious error.

The line intensities obtained by extrapolation are shown in the fifth column of Table 14.

As we see from the table, the line intensities in the liquid exceed those in the vapor over the temperature range 40 to 120°C, in agreement with the conclusions reached above. It should be noted that as the temperature rises the difference between the liquid and vapor line intensities diminishes.

In a number of papers devoted to studying the infrared [60-63] and electronic [64-67] absorption spectra, a phenomenon analogous to that described above was observed, namely, that the intensity of the electronic and infrared absorption bands rose as the substance passed from the vapor phase into liquid. The authors tried to explain the observed spectral intensity changes as being due to the presence of a so-called "effective" field in the liquids. Moreover, they used model representations (model of the "molecule-medium" system) borrowed from the polarization theory of liquid dielectrics. It is well known that such theories attempt to link macroscopic dielectric properties of the liquid (e.g., dielectric constant ε and refractive index n) with the microscopic characteristics of the constituent molecules (constant dipole moment, polarizability, etc.). With the aid of these models, the internal field E_{eff} in the liquid is expressed in terms of the field E_0 in air. In considering the phenomenon observed in the electronic and infrared absorption and Raman spectra from this point of view, we consider that the external dielectric medium surrounding the molecule in the liquid does not directly affect the probability of transitions, but only changes the magnitude of the field of the light wave acting on the molecule. In calculating E_{eff} it is usual to use the Lorentz or Onsager model.

Comparison of the spectral intensities in vapors and in solutions with the results based on the Lorentz [68] and Onsager [69] models was made in the electronic and infrared absorption spectra of the substances. The theory of the internal field was applied most coherently and successfully by B. S. Neporent and N. G. Bakhshiev [67] to the interpretation of results obtained in the absorption and emission electronic spectra of complex organic molecules. It was found in the papers mentioned that the experimental results agreed qualitatively with the Lorentz theory and closely followed results obtained on the theory of Onsager. In examining the infrared absorption spectra, a number of authors found only qualitative agreement between the experimental results and the Lorentz and Onsager theories. Let us consider our own experimental results from this point of view.

The line intensity ratio for the Raman effect in liquids and vapors will according to the Lorentz [68] model equal

$$\frac{I_{\text{liq}}}{I_{\text{v}}} = \frac{1}{n}\left(\frac{n^2 + 2}{3}\right)^2. \tag{13}$$

In Table 14 the experimental results are compared with those calculated by formula (13). In order to determine $I_{\text{liq}}/I_{\text{v}}$ we used the refractive indices of the liquids obtained by extrapolation to temperatures 120 and 80°C (superheated liquid). As seen from Table 14, the liquid-to-vapor intensity ratio with increasing liquid refractive index, which agrees qualitatively with results based on the Lorentz internal field concept. There remains, however, a considerable quantitative discrepancy. Thus on passing from the vapor to the liquid phase at 40°C the intensity of the 656 cm^{-1} line of CS_2 changes 600%, and at 80°C by 200%; according to the Lorentz theory the intensity of this line should change only 50% at these temperatures. This result is not unexpected, since Raman line intensities in liquids fall with increasing temperature (see Chapter 3, §1), and these changes cannot be explained within the framework of the model in question. Actually, according to Lorentz, the changes in intensity are caused only by changes in refractive index of the liquid. As we know, the variation of refractive index with temperature is quite small, as is the difference between the refractive indices of different liquids. According to this, the line intensity for all the liquids should change, according to formula (13), by approximately 4% on changing the temperature by 100°C. Our experiments showed that intensity changes of this order are only observed in the spectra of relatively complex molecules, such as hexylbenzene and 1,2-dicyclohexylethane. For the majority of substances, however, the changes in intensity many times exceed those calculated from formula (13). Analysis of the experimental results obtained shows that the temperature dependence of the line intensities in liquids and the liquid-to-vapor line intensity ratios cannot be explained on the basis of the Lorentz internal field concept, i.e., by the elastic polarization of the molecules only.

It is well known that Lorentz and Onsager used coarsely simplified molecular models and took no account of short-range forces, the correlation of molecules, and the kinetics of relaxation processes in liquids [70]. In view of this we must investigate not only the refractive index but also more sensitive parameters associated with the observed phenomenon and characterizing the liquid state of matter.

CHAPTER V

INVESTIGATION OF THE EFFECT
OF TEMPERATURE ON THE LINE SHAPE
OF THE RAMAN EFFECT
IN THE CONDENSED STATE OF MATTER

It was shown in Chapters 3 and 4 that the temperature dependence of Raman intensities in liquids is caused not only by the statistical redistribution of the molecules with respect to the energy levels but also by the change in intermolecular interaction with temperature. It is known that, together with intensity, one of the parameters most sensitive to intermolecular interactions in Raman scattering spectra is line shape. Our first experiments showed that the variations of these parameters with temperature are closely interlinked. A certain parallelism is observed between the temperature variation of the line intensities and the variation in the breadth of lines with a degree of polarization equal to $^6/_7$. Thus, as a rule, the intensity of the lines in the spectrum of a given molecule varies the more sharply, the sharper the variation in the breadth of the depolarized lines in the spectrum with temperature. Moreover, the sharpest temperature variations are found for molecules of the simplest structures.

By way of illustration, Fig. 7 shows the line intensity variation for liquid carbon disulfide, cyclopentane, and 1,2-dicyclohexylethane over a wide range of temperature. Figure 8 shows the temperature dependence of the breadth of the depolarized lines $\Delta \nu = 397$ cm^{-1} of carbon disulfide, $\Delta \nu = 1031$ and 1449 cm^{-1} of cyclopentane, and $\Delta \nu = 1032$ and 1445 cm^{-1} of 1,2-dicyclohexylethane. As we see from Figs. 7 and 8, the maximum temperature variations occur in the carbon disulfide line intensities; moreover the change in the breadth of the depolarized line $\Delta \nu = 397$ cm^{-1} is also the greatest. The spectral intensities of 1,2-dicyclohexylethane change least of all, and accordingly, the changes in the breadths of the depolarized lines 1032 and 1445 cm^{-1} are insignificant. The same type of law is found in the spectra of the other substances studied.

The results presented above show that a study of the temperature dependence of the line shapes may give supplementary information on the nature of the phenomenon occupying our attention. Accordingly, we made a detailed study of the effect of temperature on Raman line shapes.

§1. Temperature Dependence of Raman Line Shapes

It was shown in a paper by L. I. Sobel'man [6] that the breadth of Raman lines is connected with the kinetics of the rotary motion of the molecules in the liquid. In the case of Raman scattering, the amplitude of the scattered light is proportional to the derivative of the polarizability tensor with respect to the normal coordinate corresponding to the given oscillation. If the polarizability derivative tensor is anisotropic, the amplitude of the scattered light wave is modulated on rotation of the molecule. If the tensor is isotropic, the rotation cannot affect the amplitude. In reality the tensor contains both isotropic and anisotropic parts. The relation between these parts is determined by the degree of depolarization ρ of the lines. Hence ρ may be chosen as a criterion determining the effect of the rotary motion of the molecules on the scattering amplitude.

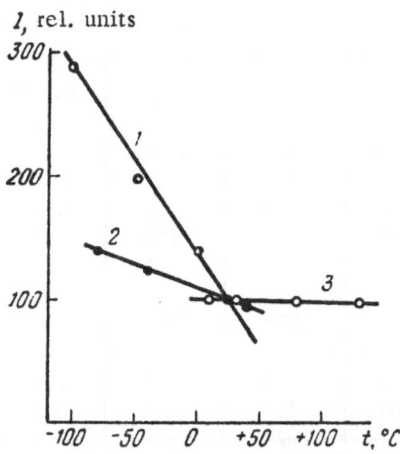

Fig. 7. Temperature dependence of Raman spectral intensities for (1) carbon disulfide, (2) cyclopentane, and (3) 1,2-dicyclohexylethane. Ordinates: intensity changes averaged over all lines measured; intensity of lines at 25°C taken as 100.

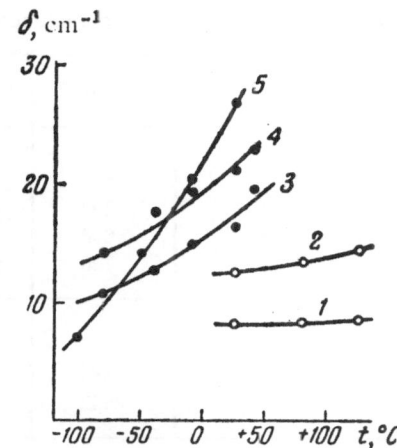

Fig. 8. Temperature dependence of the breadth of the depolarized lines. 1) 1,2-dicyclohexylethane: $\Delta \nu = 1445$ cm^{-1}; 2) $\Delta \nu = 1032$ cm^{-1}; 3) cyclopentane: $\Delta \nu = 1449$ cm^{-1}; 4) $\Delta \nu = 1031$ cm^{-1}; 5) carbon disulfide: $\Delta \nu = 397$ cm^{-1}.

In a liquid, where a chaotic Brownian type of rotatory motion is most probable, the scattering amplitude $\varepsilon(t)$ will be a random function of time if the tensor of the polarizability derivative is anisotropic ($\rho = 6/7$). Between values of this function at times t and $t + \tau$ there will be the correlation

$$\varphi(\tau) = \overline{\varepsilon(t)\varepsilon(t + \tau)}. \tag{14}$$

The function $\varphi(\tau)$ has a maximum for $\tau = 0$ and falls monotonically with increasing τ. For $\tau \to \infty$, $\varphi(\tau) \to 0$. In practice the correlation vanishes for $\tau = \tau_0$, where τ_0 has the sense of a mean time of reorientation for the molecule in the liquid, i.e., the time during which the equilibrium orientation of the molecule in the liquid remains unaltered [71]. Using the well-known connection between the spectrum of a random function and its correlation function [72], we obtain the breadth* of the spectral distribution of function $\varepsilon(t)$ as $2/\tau_0$.

On the basis of these considerations it was shown in [6] that the depolarized lines ($\rho = 6/7$) in a liquid must have a shape close to the dispersion form with breadth $\Delta \omega \geq 2/\tau_0$ (ω being the cyclical frequency). If $0 < \rho < 6/7$, we may separate out for such lines an anisotropic part $2/\tau_0$ and an isotropic part, the breadth of which does not depend on the rotatory motion. For ρ values close to zero, the line breadth will be determined only by the isotropic part and will not depend on the rotatory motion. On increasing the temperature of the liquid the mean time of reorientation of the molecules falls. Hence the breadth of the depolarized lines should increase with temperature, while the breadth of the polarized lines ($\rho = 0$) should remain constant. The results obtained in [6] are not connected with any concrete assumptions regarding the nature of the rotatory motion of the molecules. It is sufficient that this random motion should be stationary.

Experimental confirmation of the theory of L. L. Sobel'man was obtained in investigations of A. V. Rakov [9, 34, 72] and N. L. Rezaev [8, 73]. Our own investigations [11, 74] of the temperature dependence on line breadths, undertaken in order to establish their correlation with the temperature variation of intensity, also agree with the conclusions of [6]. Like the other authors, we observed that the shape of the majority of lines in the Raman effect remained close to the dispersion form, independently of the temperature. The breadth of the depolarized lines as a rule changed more sharply with temperature than that of the polarized lines. No clear connection was found between the change in line breadth with temperature and the form of the molec-

---

*Causes of line broadening not connected with rotation are not considered in the present case.

TABLE 15. Temperature Dependence of Line Breadths for Various Molecular Structures

Substance	$\Delta\nu$, cm^{-1}	ρ	δ, cm^{-1}			$\Delta\delta$, cm^{-1}
			—100° C	—50° C	25° C	
Carbon disulfide S=C=S	397	depol.	7.0	14.0	27.0	20.0
	656	0.20	0.7	0.9	0.9	0.2
	796	0.20	3.0	4.0	4.2	0.3
			—30° C	25° C	70° C	
Acetonitrile	384	depol.	7.0	9.5	13.6	5.6
	2942	pol.	5.0	5.2	5.4	0.4
			—25° C		75° C	
Benzene	607	0.88	6.6		9.2	2.6
	992	0.11	1.6		2.2	0.6
	1586	0.81	12.5		17.9	5.4
	1606	0.80	8.8		12.4	3.6
			25° C	50° C	80° C	
Hexylbenzene C—C—C—C—C—C	622	0.90	4.0	4.5	5.0	1.0
	1003	0.11	2.6	2.6	2.8	0.2
	1031	0.09	4.0	4.3	4.6	0.6
	1584	0.77	7.3	8.0	8.1	0.8
	1606	0.76	8.1	8.8	9.5	1.4
			25° C		80° C	
Cyclohexane	802	0.13	1.9		2.2	0.3
	1029	0.79	10.5		14.2	3.7
	1267	0.77	10.8		14.5	3.7
	1445	0.81	13.2		16.7	3.5
			25° C	80° C	125° C	
1,2-dicyclohexyl-ethane —C—C—	798	pol.	2.3	2.5	3.0	0.7
	1032	depol.	8.2	8.4	8.6	0.4
	1445	depol.	12.6	13.6	14.6	2.0

ular vibrations. We made a detailed study of the temperature dependence of line breadths in the spectra of a number of molecules markedly differing in dimensions and structure. The results of this are shown in Table 15.

Our measurements showed that the temperature dependence of the breadth of polarized lines did not change greatly on complicating the structure of the molecules. The temperature dependence of the breadth of depolarized lines, however, became weaker as the complication of the molecules increased. In the benzene spectrum the width of the depolarized lines increases by some 4 cm^{-1} on changing the temperature by 50°, whereas in the hexylbenzene spectrum the change in breadth is around 1 cm^{-1}. In the cyclohexane spectrum the breadth of the depolarized lines changes by approximately 3.7 cm^{-1} over a temperature range of 55°. In the 1,2-dicyclohexylethane spectrum, the breadth of the depolarized lines of the corresponding vibrations changes in all by 0.4 and 2 cm^{-1} over a temperature range of 100°.

In agreement with the principles developed in [6], the slight temperature changes in the breadth $\Delta\delta$ of the depolarized lines in the spectra of complex polyatomic molecules may be explained by the large reorientation time of these molecules. For such molecules the breadth of the lines is mainly due to processes unconnected with the rotatory motion of the molecules in the liquid (damping of vibrations under the influence of intramolecular interactions, etc.).

N. I. Rezaev [73] arrived in a somewhat different way at the conclusion regarding the insignificant effect of chaotic rotatory motion in liquids on the line breadth in the spectra of complex molecules. He compared the breadths of the lines corresponding to characteristic vibrations of a number of structurally associated molecules (benzene, toluene, ethylbenzene, n-butylbenzene, and n-hexylbenzene) at 25°C. Rezaev noted that in molecules with large dimensions (n-butylbenzene, n-hexylbenzene) the depolarized lines were of approximately the same breadth as the polarized lines, the breadth of the depolarized lines as a rule being less, and that of the depolarized lines more than in the benzene and toluene spectra. The author of [73] came to the conclusion that in the case of complex molecules (butylbenzene and hexylbenzene), the relaxation time of which is large, rotational motion has no effect on the Raman line breadths. In the benzene and toluene spectra, in view of the considerable effect of molecular reorientation, the breadth of the depolarized lines markedly exceeds that of the polarized lines.

Thus the results quoted above indicate that the reorientation of the molecules in a liquid has a considerable effect on the breadth of the depolarized Raman lines. This effect is the larger, the simpler the structure of the light-scattering molecules.

§2. Effect of Intermolecular Interaction in a Liquid on the Temperature Dependence of Raman Line Breadths

It was shown in the preceding section that the temperature variation of the breadths of depolarized lines may be explained by the influence of chaotic rotatory motion of the molecules in the liquid. Thus according to [6] the broadening of the depolarized lines is $\pi c\delta = 1/\tau_0$, where τ_0 is the mean time for the reorientation of the molecules in the liquid. Using the relationship indicated, we were able to evaluate the mean time for the reorientation of the molecules and compare with results obtained in other ways. In order to determine τ_0, it is necessary to separate out from the breadth of the depolarized line the part $\Delta\delta$ caused by reorientation of the molecules in the liquid. Remembering that in the solid phase the rotatory motion is retarded, we may find $\Delta\delta$ as the difference in line breadths at the given temperature and in the solid state.

Figure 9 shows our reorientation times at various temperatures for a number of molecules differing substantially in structure and size. As seen from the figure, the mean molecular reorientation time falls as the temperature of the liquid rises. Moreover τ_0 varies nonlinearly with temperature. As the molecular structure becomes more complicated and the size increases, the mean reorientation time rises considerably. Molecules of 1,2-dicyclohexylethane and hexylbenzene have the largest τ_0 and carbon disulfide and cyclopentane the smallest. The results obtained are also not contradictory to data published in [75-77] on a study of the relaxation of the dipole moment of liquid dielectric molecules. In particular, the authors of the papers mentioned observed a growth in relaxation time for the molecules of simple esters and ketones on increasing the chain length. A. V. Rakov established a connection between the temperature broadening of depolarized lines and the magnitude of the potential barrier of the molecule in the liquid [72].

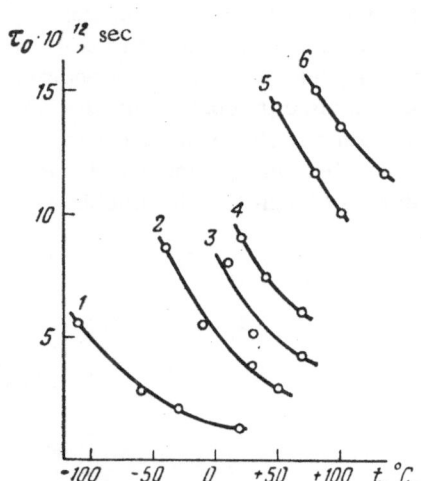

Fig. 9. Temperature dependence of mean reorientation time for various molecular structures. 1) Carbon disulfide; 2) cyclopentane; 3) benzene; 4) cyclohexane; 5) hexylbenzene; 6) 1,2-dicyclohexyl-ethane.

TABLE 16. Potential Barriers of Certain Molecules

Substance	U_1, kcal/mole from breadth	U^a, kcal/mole from viscosity	Substance	U_1, kcal/mole from breadth	U^a, cal/mole from viscosity
Carbon disulfide .	1.2	1.4	Benzene	2.4	2.4
Cyclopentane ...	1.5	1.6	Toluene	2.8	2.2
Acetone.......	1.8	1.6	Cyclohexane...	2.7	2.8

[a]The values of the viscosity coefficients for the substances under examination were taken for various temperatures from the Chemical Handbook [78].

As we know, the mean reorientation time of the molecules of the liquid is inversely proportional to the probability of the reorientation of a molecule. The reorientation probability is written in the form

$$W = Be^{-\frac{U_1}{kT}},$$

(15)

where B is a constant and U is the potential barrier of the molecule. Since the broadening of the depolarized lines approximately equals $1/\tau_0$, we may write

$$\Delta\delta = Be^{-\frac{U_1}{kT}}.$$

(16)

Hence from the temperature dependence of the breadth of the depolarized lines we may calculate the potential barrier of the molecules. Table 16 shows results of determining U_1 for certain liquids obtained in the present investigation. The mean arithmetical error in determining the potential barrier was some 7%. In the last column of the table we give the potential barrier found from the temperature dependence of viscosity.

We see from the table that the potential barriers calculated from the temperature variations of viscosity and depolarized line breadth respectively for our substances agree fairly well with each other. Thus the results presented indicate that the potential barrier of a molecule may be calculated from the temperature dependence of the depolarized Raman line breadths. Well away from phase transformation points, this parameter depends very little on temperature and characterized the value of the intermolecular interaction and the kinetics of molecular motion in the liquid phase. On the other hand, the good agreement between the potential barriers given by the temperature variations of depolarized line breadths and viscosity coefficients indicates that the potential barrier calculated from viscosity may serve as a certain criterion of the effect of temperature on the breadths of the depolarized lines in the spectra of the molecules. In agreement with our experimental data, the temperature variations of the depolarized line breadths will be greatest for liquids with the smallest potential barriers.

CHAPTER VI

EFFECT OF INTERMOLECULAR INTERACTIONS
ON THE INTENSITIES OF RAMAN SPECTRA

§1. Temperature Variation of the Spectral Intensities of Pure Liquids as a Function of the Magnitude of the Potential Barrier

The considerable amount of experimental material discussed above has shown that a very general kind of law is found for the temperature variation of line intensities in the spectrum and the breadths of depolarized lines, namely, that, as a rule, there is a sharper temperature variation in the breadths and intensities of lines ($\rho = 6/7$) for the simpler molecular structures. This law for the breadths of depolarized lines, as shown in Chapter 5, may be explained by the reorientation of molecules in a liquid. A parameter associated with the molecular structure and substantially determining the magnitude of the effect is the potential barrier of the molecule. Since the temperature variations in the spectral line intensities and the breadths of the depolarized lines are linked by a common law, it is not to be excluded that these changes in intensity may be due to processes in some measure connected with the reorientation of the molecules in a liquid. If this idea is right, we should observe a relation between the temperature variations of intensity and the magnitude of the reorientation potential barrier of the molecule. In pure liquids, we know that the potential barrier increases as the molecular structure becomes more complicated. Thus a connection between the observed effect and the potential barrier may in particular be observed on comparing the intensity changes in the spectra of molecules considerably differing in structure over the same temperature range.

Examination of the temperature dependence of intensity for the overwhelming majority of lines in the spectra of a number of structurally simple and complex molecules showed that the change in spectral line intensity with temperature decreased as the potential barrier of the molecule rose. By way of example, Table 17 shows the change in Raman spectral intensity as a function of temperature for a number of liquids differing in the height of their potential barriers.

We see from the table that, as the molecular structure becomes more complex, the potential barrier becomes greater and the temperature variation of the Raman line intensities diminishes. Thus the most liable to the temperature effect is the intensity of the Raman lines of carbon disulfide, which has the smallest potential barrier. Comparison between the temperature variations of line intensities pertaining to substances from the same class of compounds confirms this law. The intensity of the benzene lines in a temperature range of 100° varies on average by 33%, while that of hexylbenzene lines belonging to vibrations of the same group of bonds varies on average by only 6%. The intensity of cyclohexane lines, on raising the temperature 100°, changes on average by 21% ($U_1 = 2.7$ kcal/mole). The intensity changes of the 1,2-dicyclohexylethane lines lie practically within the limits of measuring error ($U_1 = 4.5$ kcal/mole). An analogous connection between the temperature variation of Raman line intensities and potential barrier is also found for the other substances studied (cyclopentane and 1,2-dicyclopentylethane; dipropenyl and alloocimene). It should be noted that the law in question is as a rule not broken even when we compare substances not belonging to the same class of compounds.

TABLE 17. Changes in Raman Line Intensities with Temperature as a Function of the Potential Barrier of the Substance

Substance	ΔI_{av}, % for 100°C		Substance	ΔI_{av}, % for 100°C	U, kcal/mole
Carbon disulfide S=C=S	150	1.2	Cyclohexane	21	2.7
Cyclopentane	40	1.5	1,2-dicyclohexylethane —C—C—	2	4.5
1,2-dicyclopentylethane —C—C—	20	3.8	Dipropenyl C—C=C—C=C—C	35	2.4
Benzene	33	2.4	Alloocimene C—C=C—C—C=C—C=C—C	15	4.0
Hexylbenzene —C—C—C—C—C—C	6	4.2			

§2. Study of the Temperature Dependence of the Raman Spectral Intensities of Solutions

It was established in the preceding section that in pure liquids the temperature variations of spectral intensities depended on the size of the potential barrier. From the point of view of confirming this relationship, it was of interest to carry out analogous investigations with solutions, artificially varying the potential barrier of the molecule. In accordance with the principles of Ya. I. Frenkel' [70] we may assume that for nondipolar, chemically noninteracting molecules at low solution concentrations (3 to 8%) the potential barrier of a molecule of the dissolved substance is to a considerable extent determined by the solvent. Of course, in view of the peculiarities of the structure of solutions, we can hardly expect complete agreement between the temperature variations of the line intensities of different substances in solvents with the same potential barriers. But the law governing the potential barrier should to a certain extent appear in solutions also.

As solvents we selected liquids with various potential barriers. The magnitude of the potential barrier of the solvent was previously determined either from the temperature dependence of the depolarized line breadths or from that of the viscosity coefficient.* The results of measuring the line intensities of substances in solution at various temperatures are given in Table 18. The first column shows the solvent and its potential barrier U; the last two give the relative changes (in 50°C) of the line intensities of the pure substance and the line intensity changes of the same substance in solution.

As we can see from the table, the temperature variations of line intensity for a particular substance increase on solution in a liquid with a smaller potential barrier. In the present case, the solvent with the lowest potential barrier is carbon disulfide (U = 1.2 kcal/mole). Whereas in pure carbon tetrachloride (U = 2.6 kcal/mole) the line intensity changes over a temperature range of 50° are approximately 15%, for a 5% solution of carbon tetrachloride in a carbon disulfide these changes reach 70% over the same temperature range.

* The potential barrier of triacetin was calculated from the temperature dependence of viscosity taken from the dissertation of T. V. Velichkina [79].

TABLE 18. Temperature Dependence of Raman Line Intensities for a Number of Substances in Solutions

Solvent and its potential barrier U, kcal/mole	ΔI, % for 50°C	
	Pure substance	Dissolved substance
Carbon disulfide, $\Delta \nu = 656$ cm^{-1}, c = 3%, U = 1.2 kcal/mole		
Benzene, 2.4	70	15
Triacetin, 6.0	70	20
Acetone, 1.8	70	68
Carbon tetrachloride, 2.6	70	50
Benzene, $\Delta \nu = 992$ cm^{-1}, c = 3%, U = 2.4 kcal/mole		
Carbon disulfide, 1.2	25	40
Carbon tetrachloride, 2.6	25	10
Triacetin, 6.0	25	10
Carbon tetrachloride, $\Delta \nu = 314$ cm^{-1}, c = 5%, U = 2.6 kcal/mole		
Carbon disulfide, 1.2	15	76
Triacetin, 6.0	15	4
Carbon tetrachloride, $\Delta \nu = 459$ cm^{-1}, c = 5%		
Carbon disulfide, 1.2	11	72
Triacetin, 6.0	11	5

In a 3% solution of benzene (U = 2.4 kcal/mole) in carbon disulfide, the change in the intensity of the 992 cm^{-1} benzene line with temperature is also considerably greater than that for the pure substance, although the difference is less marked than in the case of carbon tetrachloride. On the other hand, on dissolving the substances in question in liquids with a higher potential barrier, for example triacetin (U = 6.0 kcal/mole), the variation of the spectral intensity with temperature as a rule diminishes. Thus the intensity of the 656 cm^{-1} line of pure carbon disulfide changes by approximately 70% in a temperature range of 50°. On dissolving the carbon disulfide in triacetin, the temperature variation of the intensity of the same line is no greater than 20%. Analogous laws are found for solutions of other substances.

We studied solutions of benzene and carbon tetrachloride in triacetin not only in the liquid phase but also in glass form. It should be noted that these solutions in the glassy state remained completely transparent. This is extremely important in quantitative measurements.

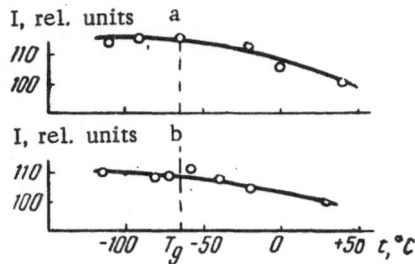

Fig. 10. Temperature dependence of Raman line intensity. a) Benzene dissolved in triacetin (concentration 5%), $\Delta \gamma = 992$ cm^{-1}; b) carbon tetrachloride dissolved in triacetin (concentration 5%), $\Delta \gamma = 459$ cm^{-1}.

According to the principles of the theory of the liquid state [70], the potential barrier of a molecule rises sharply on passing from the liquid phase into the glassy state. Hence, if the concept of a link between the temperature variation of intensity and the height of the potential barrier is correct, then in glasses the anomalous temperature dependence of intensity should hardly be found at all. The temperature dependence of the line intensities of 5% solutions of benzene and carbon tetrachloride in triacetin are shown in Fig. 10. As seen from the figure, the intensity of the lines in liquid solutions, as usual, increases on lowering the temperature. In glasses, however, the line intensities remain practically constant over the whole temperature range.

Thus on the whole the results presented quite well illustrate the connection between the temperature variations of Raman line intensities with the size of the potential barrier of the liquid. The existence of such a connection indicates that the reorientation of molecules in the condensed phase must have a substantial effect on the intensity of the Raman spectra.

CHAPTER VII

DISCUSSION OF RESULTS

Our experimental investigations have shown that the anomalous temperature dependence of Raman spectral intensities for liquids cannot be explained by existing theory. Up to the present, unfortunately, there has not been even a qualitative model enabling us to picture the mechanism of this phenomenon. The fact that an analogous situation exists in infrared absorption spectra calls for our attention. As we know from theory, the intensity of the infrared absorption bands is independent of temperature [80]. Recently a number of authors [81-90], experimentally investigating the temperature dependence of infrared absorption spectra, have observed laws similar to those found in Raman spectra, namely, that the temperature variations of the infrared absorption band intensities of liquids are opposite in sign to those expected from theory. In gases there is qualitative agreement between theory and experiment.

At the present time it is hard to establish an exact quantitative connection between the observed phenomena in the visible and infrared parts of the spectrum, owing to the absence of enough experimental material on simultaneous studies of the infrared absorption spectra and the Raman effect for the same substances. The similar signs of the effect in both liquids and gases in each case, however, leads us to think that the phenomena observed have the same nature. In view of this, in discussing the results of the present work and results recently obtained by other authors on the temperature dependence of Raman spectral intensities, we shall also touch on a number of papers devoted to studying the temperature dependence of infrared absorption band intensities.

Roesler [26] discussed the proposition that the fall in intensity of vibrational spectral lines as temperature increases is the result of incomplete account being taken of the wings of the line by experimentalists. The determination of integral intensities from the line maxima when working with a constant apparatus slit width, and also the lack of proper allowance for the wings when measuring the integral intensities from the area bounded by the line contour, may indeed lead to the errors indicated in [26]. The errors should be the larger, the more the line broadens on raising the temperature. Our experiments showed, however, that the fall in Raman line intensity for liquids on raising the temperature cannot be explained by errors arising from the neglect of the wings. This may be illustrated from the example of the polarized lines, the breadth of which is practically independent of temperature (see Chapter 5, §1). Table 19 shows the breadth and intensity of the polarized lines as functions of temperature for a number of substances. We see from the table that the line widths of all substances considered change little on raising the temperature; the intensity of heptene-1 and diallyl, however, falls by some 1.5 times, and that of carbon disulfide by 3 times. Hence the observed phenomenon is quite genuine, and cannot be explained by the experimental errors discussed.

Certain authors have suggested [91, 92] that the anomalous temperature dependence in Raman line intensities is caused by changes in the electronic absorption spectra under the influence of temperature. For example, the fall in Raman line intensity with increasing temperature might be connected with the simultaneous shift of the electronic absorption bands into the long-wave region [92]. Such a displacement of the electronic absorption bands was actually observed for a series of organic liquids by N. G. Bakhshiev [93],

TABLE 19. Changes in Breadth and Intensity of Raman Lines as Functions of Temperature

Substance	$\Delta\nu$, cm^{-1}	α	t, °C	σ, cm^{-1}	$\Delta\sigma$, cm^{-1}	I_{t_1}/I_{t_2}	Substance	$\Delta\nu$, cm^{-1}	α	t, °C	σ, cm^{-1}	$\Delta\sigma$, cm^{-1}	I_{t_1}/I_{t_2}
Heptene-1	1642	0.15	—10 / 90	4.0 / 5.5	1.5	1.35	Carbon disulfide	656	0.20	—110 / 25	0.7 / 0.8	0.1	2.9
Diallyl	1642	0.12	—70 / 55	6.0 / 6.2	0.2	1.40		796	0.20	—110 / 25	4.1 / 4.5	0.4	2.5

L. G. Pikulik [94], and by V. V. Zelinskii and V. N. Kolobkov [95]. * It is not impossible that the displacement of the electron absorption bands may influence the temperature variation of Raman line intensities near the resonance region. It is hard to picture, however, that for nondipolar substances, for which the band displacement is as a rule, according to [93], fairly small, and the absorption bands are situated in the far ultraviolet (for example, in the case of benzene, cyclopentane, and carbon tetrachloride), the effect in question can determine the character of the changes in Raman-effect line intensities with temperature. The most practicable method of checking the connection between the temperature variations of Raman intensities and the position of the electronic absorption band lies in studying the temperature dependence of the Raman line intensities as a function of the wavelength of the exciting light.

We determined the temperature variations of the intensity of the 656 cm^{-1} line of carbon disulfide for different exciting lines Hg 4358 and 5461 A. The absorption coefficient of carbon disulfide in the visible spectrum was determined earlier (see Chapter 3, §2). Allowance for the absorption of the exciting and scattered light for various temperatures was made by means of the formula derived in [54]. The experimental results obtained showed that there was no marked difference in the temperature variations of the intensity of line 656 cm^{-1} obtained with the different excitations.

Measurements of the temperature dependence of the spectra of n-nitrotoluene and n-nitrobenzene in solutions excited by the 4358 and 5461 A Hg lines were carried out by Ya. S. Bobovich and V. M. Pivovarov [91]. The authors came to the conclusion that the frequency of the exciting light, i.e., the closeness to the condition of resonance, did not affect the temperature dependence of the Raman line intensities in liquids. Thus the anomalous temperature dependence of the Raman intensity in liquids can scarcely be caused by the displacement of the electronic absorption bands under the influence of temperature. The anomalous temperature dependence of Raman line intensities may also be connected with a change in the value of the matrix elements of the electron transitions, i.e., with a change in the integral intensity of the electronic absorption bands. Published data [96] on the temperature dependence of the integral intensity of the electronic absorption spectra are far from complete, and in a number of cases temperature variations differing in sign have been observed. The nature of these changes has moreover so far not been finally explained.

In view of what has been said, there is at present no foundation for saying that the observed anomalous temperature variation of Raman spectral intensities is caused by the changes in electronic absorption spectra mentioned.

The authors of [17-19] have tried to explain the temperature dependence of the intensity of vibrational spectra in the liquid phase by the mechanical anharmonicity of intramolecular vibrations. This proposition can hardly be valid, since the experimental measurements in the majority of investigations were carried out at fairly low temperatures ($h\omega/kT < 1$), for which the population of the excited vibrational states is not large. The effect of mechanical anharmonicity may be substantial at higher temperatures. Calculations made by a

---

* We shall not stop to consider the various interpretations of the observed displacement of the electronic absorption bands, since at present there is no single view on this, and the question is under constant discussion.

number of authors [81, 97] showed that making allowance for mechanical anharmonicity leads to an increase in line intensities with temperature, i.e., to an effect opposite in sign to that observed in liquids.

Theoretical calculation of the effects of electro-optical anharmonicity in the temperature dependence of the Raman line intensities of polyatomic molecules is beset by considerable difficulties. On the basis of semiclassical theory [3], however, we may conclude that, if the temperature dependence of the spectral intensity is associated with a change of electro-optical anharmonicity, then this should appear most strongly in the overtones. Ya. S. Bobovich and V. M. Pivovarov [91] showed that in the spectra of liquid carbon tetrachloride and chloroform (remembering the temperature factors connected with the population of vibrational levels) no marked difference is found between the temperature variations of intensity in the fundamental tones and overtones.

Our investigations also showed that in the carbon disulfide spectrum the temperature variations of the intensity of the fundamental tone 397 cm^{-1} differ little from those of the intensity of the overtone 796 cm^{-1}.

Taking account of all that has been said, we may fairly assume that neither mechanical nor electro-optical anharmonicities determine the temperature dependence of the spectral intensities.

The experimental results obtained in the present investigation when studying the spectra of liquids and vapors have enabled us to come to the conclusion that the anomalous temperature dependence of Raman line intensities for liquids is caused by intermolecular interactions. This point of view is at present shared by the majority of authors studying the temperature dependence of Raman [93] and infrared absorption line intensities [81-90]. The conclusions of different authors on the mechanism of the phenomenon are, however, different and contradictory.

In recent papers published by Ya. S. Bobovich and T. P. Tulub [92, 98], it was suggested that temperature variations in Raman line intensities were connected with a change in the internal field in the liquid. The authors give no specific model for this, only indicating that the macroscopic parameter determining the effect is the refractive index of the liquid. In fact the Raman spectral intensities, the refractive index, and the density of the substance do change in the same sense when the temperature of the liquid is varied. The authors of [92, 98] came to the conclusion that the changes in Raman spectral intensities with temperature were the larger, the larger the derivative dn/dt (n = refractive index of the liquid, t = temperature). Some of the experimental data presented in [98] to illustrate this relationship are not, however, quite convincing. In particular, the authors observed a growth in line intensity with temperature in a solution of zinc chloride in water, i.e., an effect of opposite sign to that generally found in liquids. They suggested that this temperature dependence of intensity was caused by the low temperature coefficient of the refractive index of the solvent (9 · 10^{-5}). But the experiment in question admits of a whole series of other interpretations, zinc chloride being an electrolyte. The strong effect of the electric fields of the ions on the electro-optical parameters of the molecules was observed in investigations by S. Mints, Z. Kentskii, and S. Osetskii [99, 100], who studied solutions of zinc chloride in methyl alcohol, and also aqueous solutions of uranyl nitrate. It is not impossible that the effects observed by Ya. S. Bobovich and T. P. Tulub are connected with a change in the coefficient of electrolytic dissociation, which, as we know, depends on temperature and concentration. In view of this we can scarcely compare the results of [98] with the data for pure liquids.

We considered the results obtained in the present work from this point of view, i.e., the temperature variation of the spectral intensities of liquids with different values of dn/dt. Table 20 shows the changes in spectral intensity for 100° together with the temperature coefficient of the refractive index for several liquids. As seen from the table, in comparing different liquids it is quite hard to establish a quantitative connection between these quantities. On the other hand, if we compare liquids with molecules of similar structure (benzene and toluene, carbon tetrachloride and chloroform), a relationship of this kind does appear. An analogous relationship in infrared absorption spectra with respect to dρ/dt (ρ = density of the substance) was found by M. P. Lisitsa and colleagues [84-90].

M. P. Lisitsa, V. N. Malinko, and L. N. Khalimonova [85] suggested that the fall in the infrared absorption band intensities with increasing temperature was caused by a change in the intermolecular distances and the associated fall in the strong field exerted on an absorbed molecule by its neighbors. Corresponding

TABLE 20. Changes in the Spectral Intensities with Temperature for a Number of
Liquid as a Function of dn/dt

Substance	$\Delta I, \%$ for $100°$ C	$dn/dt \cdot 10^{-5}$	Substance	$\Delta I, \%$ for $100°$ C	$dn/dt \cdot 10^{-5}$
Carbon disulfide . .	150	78	Toluene	20	56
Acetone	40	50	Chloroform	28	59
Cyclopentane	40	56	Carbon		
Benzene	33	64	tetrachloride.	25	55
			Cyclohexane	20	56

to this there should be a fall in the mean induced dipole moment of the molecule and its derivative with respect to the normal coordinate, i.e., the intensity of the absorption band. It is not to be excluded that such effects may take place. The mechanism of the pher. omenon observed can, however, hardly be reduced to merely a change in the distance between molecules. In particular, from this point of view it is hard to explain the similar character in the intensity changes with temperature in liquids and certain crystals [22-24, 92, 101-103], the density of which is practically independent of temperature.

Recently there have appeared many papers devoted to a study of the intensities of molecular spectra in the condensed state of matter, in which the experimental material is discussed from the point of view of the behavior of the internal field in a liquid [60-67, 92, 101-103]. Moreover, in order to explain E_{eff} (the field acting on the molecule), use is made of a model representation borrowed from the theory of polarization of liquid dielectrics. Usually the Lorentz [68] or Onsager [69] models are used. As already mentioned above (see Chapter 3, §3), in considering from this point of view phenomena observed in electronic and infrared absorption as well as Raman spectra, it is supposed that the external dielectric medium surrounding a molecule in a liquid has no direct effect on the probability of transitions, but only changes the value of the light wave field acting on it. The intensity of the Raman effect will in this case be proportional to E_{eff}^2.

The Lorentz model was used by Person [104] to explain the anomalous temperature dependence of the infrared absorption band intensities of benzene and cyclohexane. The author found quite good agreement between the results obtained and the relationship following from the Lorentz formula [see Chapter 4, §2, formula (13)]. In papers by other authors [30, 81, 84] the temperature variations of molecular spectral intensities considerably exceeded those calculated from formula (13). Our own experiments showed that only for spectra of complex polyatomic molecules were the intensity variations of approximately the order (5 to 6%) given by the model in question. For the majority of our liquids, the temperature dependence of the spectral intensities was considerably greater than would follow from the Lorentz model.

From the point of view of explaining the role of intermolecular interactions of a dielectric character in the observed phenomenon, it was of interest to compare our data on the temperature dependence of the spectral intensities of solutions with results based on the Onsager model . As we know, according to Onsager, the field acting on a molecule in a liquid is expressed by the formula

$$\bar{E}_{eff} = \frac{1}{\sqrt{n}} \frac{3n^3}{(2n^2+1)} \frac{1}{1 - \frac{\alpha}{r^3} \frac{2n^2-2}{2n^2+1}} \bar{E}_b, \qquad (17)$$

where n is the refractive index of the solvent, α the polarizability of a molecule of dissolved substance, r the radius of the cavity which the dissolved molecule occupies, and E_b the electric field intensity of the exciting light.

In this model, in contrast to that of Lorentz, account is taken not only of the polarization of the solvent by the light wave but also the polarization by the dipole induced in the molecule under investigation by the

TABLE 21. Temperature Variation in the Intensity of Lines Obtained Experimentally and by Calculation from Formula (17)

Solvent	ΔI, % for 50°	
	Experiment	Calculation
Benzene, c = 3%		
Carbon disulfide........	40	22
Carbon tetrachloride.....	10	16
Carbon tetrachloride, c = 5%		
Carbon disulfide........	70	7

same wave. It must be noted that in calculating E_{eff} from formula (17) there arise difficulties associated with the choice of cavity radius r, i.e., whether it should be the structural value, the kinetic value, or that determined from the number of molecules in unit volume. This is extremely important, since expression (17) is most sensitive to change in the term α/r^3. Certain authors [93, 105-107] considered that the cavity radius was equal to the structural radius of the molecule. This may well be valid for complex polyatomic molecules [93, 107], for which the radii in question are, as a rule, close together. However, for molecules of comparatively simple structure, these radii may be considerably different. Thus, for example, the structural radius for the benzene molecules is approximately 1.89 A, the kinetic radius 4 A, and the radius obtained from the number of molecules per unit volume 5.2 A. Table 21 shows the temperature variation of line intensities for solutions of benzene and carbon tetrachloride, both as determined from our experiments and as calculated on the Onsager model; r is taken as the structural radius of the molecule. As seen from the table, in this case the experimental data deviate from the theoretical. If we use the radius obtained from the number of molecules per unit volume, or the kinetic radius, the discrepancy increases.

Thus we may conclude from what has been said that, for our pure liquids and solutions consisting of molecules fairly simple in structure, having low potential barriers (U < 4 kcal/mole), the temperature dependence of spectral intensities cannot be explained by the influence of intermolecular interactions of a dielectric character based on the Lorentz or Onsager models. It is not impossible that the inapplicability of these models in the present case is connected with their omissions: both the Lorentz and Onsager models fail to allow for short-range forces and the correlation of the molecules in the liquid. In particular, in the Onsager model the binding between a molecule and its neighbors is considered only as a reactive field, which, being parallel to the direction of the dipole moment of the molecules under consideration, does not alter its rotation at all.

In the theory of liquid dielectrics, of course, there exist more complete models, in which some of the failings in question are remedied [108, 109]. These models, however, apart from elastic electronic polarization, consider mainly the orientation polarization of the dielectric, which plays no great part in the region of optical frequencies. These can thus hardly provide anything new in explaining the phenomenon of present interest. Furthermore, the models in question take no account of the kinetics of relaxation processes in the liquid.

As our experimental studies have shown, the reorientation of molecules in a liquid has a considerable effect on the shape and integral intensity of Raman lines. This effect is especially strong in the spectra of molecules with small potential barriers. The reorientation of molecules in a liquid has practically no effect on the Raman spectrum if the potential barrier exceeds 4 kcal/mole. Polyatomic "complex" molecules (hexylbenzene, 1,2-dicyclohexylethane, etc.) have such a barrier. It is not to be excluded that the temperature dependence of intensity, and the change in intensity on passing from one state of aggregation into another, may be explained for such molecules by intermolecular interactions of a dielectric character. In the case in which the dynamics of the molecules in the liquid are important, the statistical relationships of Lorentz and Onsager cannot completely describe the phenomena observed. Unfortunately, up to the present time there has been no sufficiently complete theory of the liquid state, the formation of such a theory being beset by considerable difficulties. In view of this, in interpreting the experimental results obtained, we must confine ourselves to merely qualitative considerations.

The totality of information on the structure of liquids obtained by various methods, especially by means of x-ray studies, shows that there exist correlating interactions between molecules in the liquid state, in contrast to the gaseous state. In this case, even if the direct interaction is small, there may be a connection via the radiation field, i.e., each molecule may be situated in the field of a neighboring molecule. Thus, when light interacts with matter, the absorbing, radiating, or light scattering molecules cannot be considered as absolutely independent of one another. As we know, the theory of the intensity of the Raman effect was developed for noninteracting molecules. In this case the phases of the scattered light waves very arbitrarily from one scattering center to another, and the intensity of the Raman scattering is directly proportional to the number of scattering particles per unit volume. This assertion is completely valid in the case of rarefied gases. It is not impossible, however, that in liquids the vibrations of the scattering centers must be treated as coupled. *

The scattering of light in a coupled system will be rather different. For fairly strong coupling and a wavelength exceeding the dimensions of the system, the vibrations of the scattering centers may be partly coherent. If this model is correct for liquids, then we can explain the experimental results observed. In this case we may expect a rise in Raman intensity as the substance passes from the vapor to the liquid state. It is extremely probable that, on increasing the temperature of the liquid, as a result of the increase in molecular thermal motion, the partial coherence will fall, and the Raman intensity will diminish. Stronger changes in intensity must evidently be expected for liquids with low potential barriers. From this point of view we can explain the remaining laws noted in the present investigation.

It should be mentioned that we did also obtain experimental data which can be regarded as a certain confirmation of the explanation offered above. We made measurements of the indicatrix of the Raman effect for liquid carbon disulfide, benzene, carbon tetrachloride, and toluene. Experiment showed that, for all the liquids without exception, a phenomenon analogous to the Mee effect were observed, namely, that the scattering along the direction of the exciting light was considerably greater than in the opposite direction. This leads us to the conclusion that Raman scattering in liquids is apparently in part coherent.

The above ideas regarding the causes of the phenomena observed in liquids will not allow us to make any quantitative comparisons, since the theoretical calculations which would have to be made are far from trivial. It is clear, however, that in order to solve the problem of the temperature dependence of Raman intensities in liquids we must develop a theory which will take due account of short-range forces and the kinetics of relaxation processes in the condensed state of matter.

The author expresses thanks to Doctor of Physicomathematical Science Professor P. A. Bazhulin for direction and discussion of the results of this work.

---

* The idea of the coupled motion of interacting molecules in liquids was discussed by M. V. Vol'kenshtein [110] in 1941.

LITERATURE CITED

1. G. S. Landsberg and L. I. Mandelstam, Z. Physik 60:364 (1930); I. E. Tamm, Z. Physik 60:345 (1930).
2. G. Plachek, Rayleigh Scattering and the Raman Effect, Khar'kov, ONTIU (1935).
3. M. V. Vol'kenshtein, M. A. El'yashevich, and G. I. Stepanov, Vibrations of Molecules, Vol. 2, Gostekhizdat (1949).
4. V. Heitler, Quantum Theory of Radiation [Russian translation] IL (1956).
5. G. Dirac, Fundamentals of Quantum Mechanics [Russian translation] GIZ (1932).
6. L I Sobel'man, Izv. Akad. Nauk SSSR, Ser. Fiz. 17:5 (1953).
7. I. I Sobel'man, Tr. Fiz. Inst. Akad. Nauk 9:315 (1958).
8. N. I. Rezaev, Material of the Tenth All-Union Conference on Spectroscopy, Vol. 1, L'vov (1957), p. 230.
9. A. V. Rakov, Material of the Tenth All-Union Conference on Spectroscopy, Vol. 1, L'vov (1957), p. 229.
10. G. V. Mikhailov, Material of the Tenth All-Union Conference on Spectroscopy, Vol. 1, L'vov (1957), p. 227.
11. P. A. Bazhulin and A. I. Sokolovskaya, Material of the Tenth All-Union Conference on Spectroscopy, Vol. 1, L'vov (1957), p. 225.
12. K. S. Krishnan, Nature 27:650 (1928).
13. A. Dadien and K. Kohlrausch, Phys. Zs. 30:384 (1929).
14. F. G. Brickwede and M. F. Peters, Phys. Rev. 33:116 (1929).
15. L. S. Ornstein and J. J. Went, Proc. Amst. Acad. Sci. 7A:196 (1932).
16. L. S. Ornstein and J. J. Went, Physica 2:503 (1935).
17. R. Ananthakrishnan, Proc. Ind. Acad. Sci. 7A:196 (1938).
18. J. Venkatesvarlu, Proc. Ind. Acad. Sci. 19:111 (1944).
19. J. Venkatesvarlu, Proc. Ind. Acad. Sci. 21:24 (1945).
20. J. Venkatesvarlu, Current Sci. 16:1 (1947).
21. J. Venkatesvarlu, Current Sci. 16:15 (1947).
22. J. Venkatesvarlu, Proc. Ind. Acad. Sci. 13:64 (1941).
23. J. Venkatesvarlu, Proc. Ind. Acad. Sci. 14:529 (1941).
24. J. Venkatesvarlu, Proc. Ind. Acad. Sci. 16:45 (1942).
25. L. M. Fishkova, Dokl. Akad. Nauk SSSR 75:523 (1950).
26. F. G. Roesler, Acta Phys. Austriaca 5:477 (1962).
27. Ya. S. Bobovich and D. K. Arkhipenko, Dokl. Akad. Nauk SSSR, 86:247 (1952).
28. Ya. S. Bobovich and V. M. Pivovarov, Dokl. Akad. Nauk SSSR 97(5):801 (1954).
29. Ya. S. Bobovich, Dokl. Akad. Nauk SSSR 98:158, 39 (1954).
30. Ya. S. Bobovich and V. M. Pivovarov, Physics Collection of L'vov State University 3(8):223 (1957).
31. V. M. Chulanovskii, M. P. Burgova, and A. N. Mironova, Izv. Akad. Nauk SSSR, Ser. Fiz. 12:560 (1948).
32. V. M. Chulanovskii, M. P. Burgova, and A. N. Mironova, Izv. Akad. Nauk SSSR, Ser. Fiz. 14:406 (1950).
33. Kh. E. Sterin, Tr. Fiz. Inst. Akad. Nauk 9:15 (1958).
34. P. A. Bazhulin and A. V. Rakov, Dokl. Akad. Nauk SSSR 105:54 (1955).
35. H. L. Welsh, P. E. Pasheer, and B. P. Stoicheff, Can. J. Phys. 30(2):99 (1952).
36. P. P. Shorygin and L. Z. Osityanskaya, Dokl. Akad. Nauk SSSR 98:51 (1954); P. P. Shorygin and A. Kh. Shalimov, Dokl. Akad. Nauk SSSR 81:1031 (1951).

37. M. M. Sushchinskii, Tr. Fiz. Inst. Akad. Nauk 12:54 (1960).

38. L. A. Woodward and J. H. B. George, Nature 167:193 (1951); Proc. Phys. Soc. B64:780 (1951).

39. G. G. Slyusarev, Geometrical Optics, Izd. AN SSSR (1954), p. 138.

40. S. G. Rautian, Zh. Eksperim. i Teor. Fiz. 27:625 (1954).

41. V. M. Pivovarov and Ya. S. Bobovich, Opt. i Spektroskopiya 3:134 (1957).

42. H. J. Bernstein and G. Allen, J. Opt. Soc. Amer 45:237 (1955).

43. A. L Sokolovskaya and S. G. Rautian, Opt. i Spektroskopiya 6(1):51 (1959).

44. D. G. Rea, J. Opt. Soc. Amer. 49(1):90-101 (1959).

45. G. S. Landsberg, P. A. Bazhulin, and M. M. Suchchinskii, Fundamental Parameters of the Raman Spectra of Hydrocarbons, Izd. AN SSSR (1956).

46. P. A. Bazhulin, A. I. Sokolovskaya, S. G. Rautian, and M. M. Sushchinskii, Izv. Akad. Nauk SSSR, Ser. Fiz. 6:678 (1954).

47. P. A. Bazhulin, S. G. Rautian, A. L Sokolovskaya, and M. M. Sushchinskii, Zh. Eksperim. i Teor. Fiz. 29:822 (1955).

48. S. G. Rautian, Usp. Fiz. Nauk 66(3):475 (1958).

49. G. G. Petrash, Opt. i Spektroskopiya 6:792 (1959).

50. M. L. Sosinskii, Izv. Akad. Nauk SSSR, Ser. Fiz. 17:621 (1953).

51. A. M. Bogomolov, Opt. i Spektrosckopiya 9:311 (1960).

52. A. L Sokolovskaya and P. A. Bazhulin, Opt. i Spektroskopiya 8:394 (1960).

53. A. L Sokolovskaya and Z. Kercki, Bull. Acad. Polon. Sci. Ser. Sci. Chim., Geol., Geograph. 6(2):138 (1958).

54. G. Michel, Spectrochim. Acta 12(4):400-402 (1958).

55. Von J. Behringer, Z. Elektrochim. 62(5):544-567 (1958).

56. A. L Sokolovskaya, Opt. i Spektroskopiya 9(5):582 (1960).

57. A. L Sokolovskaya, Opt. i Spektroskopiya (in press).

58. Ya. S. Bobovich and V. L Pivovarov, Opt. i Spektroskopiya 6:249 (1959).

59. S. P. Stoicheff, Can. J. Phys. 32:330 (1954).

60. E. E. Ferguson, J. Chem. Phys. 30(4):1059 (1959).

61. C. La Lau, Preprint of paper to be read at the Institute of Petroleum Group Conference on Molecular Spectroscopy, February, 1958.

62. W. B. Person, J. Chem. Phys. 28(2):319 (1958).

63. S. R. Polo and M. K. Wilson, J. Chem. Phys. 23:2376 (1955).

64. E. Rabinowitch and A. W. Wood, Trans. Faraday Soc. 32:540 (1936).

65. C. Child and O. Walker, Trans. Faraday Soc. 34:1066 (1938).

66. J. Platt, G. Pusoff, and H. Klevens, J. Chem. Phys. 11:535 (1943); J. Platt and H. Klevens, Rev. Mod. Phys. 16:182 (1944); J. Platt and H. Klevens, Chem. Rev. 41:301 (1947).

67. B. S. Neporent and N. G. Bakhshiev, Opt. i Spektroskopiya 5:634 (1958).

68. H. Lorentz, Theory of Electrons [Russian translation] ONTI (1935).

69. L. Onsager, J. Amer. Chem. Soc. 1:58 (1936).

70. Ya. L Frenkel', Collection of Selected Works, Vol. 3, Izv. Akad. Nauk SSSR (1959).

71. V. L Danilov, Structure and Physical Properties of Matter in the Liquid State, Kiev, Izd. AN Ukr. SSR (1954).

72. A. V. Rakov, Opt. i Spektroskopiya 7:202 (1959).

73. N. I. Rezaev, Vestn. Mosk. Univ. (2):145 (1957); Opt. i Spektroskopiya 5:5 (1958).

74. P. A. Bazhulin and A. L Sokolovskaya, Collection of Works in Memory of Academician G. S. Landsberg, Izv. Akad. Nauk SSSR (1958), p. 56.

75. F. Charles and F. Smythe, Izv. Akad. Nauk SSSR, Ser. Fiz. 24(1):25 (1960).

76. G. B. Rathmann, A. J. Curtis, P. L. Mc Geer, and C. P. Smyth, J. Chem. Phys. 25:413 (1956).

77. J. H. Calderwood and C. P. Smyth, J. Amer. Chem. Soc. 78:1295 (1956).

78. Chemical Handbook, Vol. I, Goskhimizdat (1951).

79. T. V. Velichkina, Tr. Fiz. Inst. Akad. Nauk 9:61 (1958).

80. B. L Stepanov and V. P. Gribkovskii, Dokl. Akad. Nauk SSSR 121:446 (1958).

81. P. A. Bazhulin and V. N. Smirnov, Opt. i Spektroskopiya 6:745 (1959).

82. V. N. Smirnov and P. A. Bazhulin, Opt. i Spektroskopiya 7:193 (1959).

83. V. N. Smirnov, Opt. i Spektroskopiya 7:472 (1959).

84. M. P. Lisitsa and V. N. Malinko, Izv. Akad. Nauk Ser. Fiz. 22(9):1117 (1958).

85. M. P. Lisitsa, V. N. Malinko, and I. N. Khalimonova, Opt. i Spektroskopiya 7:638 (1959).

86. M. P. Lisitsa and Yu. P. Pyashchenko, Opt. i Spektroskopiya 9:742 (1960).

87. M. P. Lisitsa and Yu. P. Pyashchenko, Opt. i Spektroskopiya 9:438 (1960).

88. M. P. Lisitsa and Yu. P. Pyashchenko, Opt. i Spektroskopiya 9:188 (1960).

89. M. P. Lisitsa and Yu. P. Pyashchenko, Opt. i Spektroskopiya 10:157 (1961).

90. M. P. Lisitsa and V. N. Malinko, Opt. i Spektroskopiya 6:694 (1959).

91. Ya. S. Bobovich and B. M. Pivovarov, Opt. i Spektroskopiya 3:227 (1957).

92. Ya. S. Bobovich and T. P. Tulub, Opt. i Spektroskopiya 9:745 (1960).

93. N. G. Bakhshiev, Opt. i Spektroskopiya 6:250 (1959); 7:52 (1959).

94. L. G. Pikulik, Physical Problems in Spectroscopy, Material of the Thirteenth Conference, Vol. 1, Izd. AN SSSR 297 (1962), p. 297.

95. V. V. Zelinskii and V. P. Kolobkov, Opt. i Spektroskopiya 5:423 (1958).

96. V. L. Levshin, Photoluminescence of Solids and Liquids, Izd. AN SSSR (1951).

97. M. P. Lisitsa and V. L. Strizhevskii, Opt. i Spektroskopiya 10:48 (1961).

98. T. P. Tulub and Ya. S. Bobovich, Opt. i Spektroskopiya 9:669 (1960).

99. S. Mints and Z. Kéntskii, Izv. Akad. Nauk SSSR, Ser. Fiz. 22(10):1182 (1959).

100. S. Mints and S. Osetskii, Izv. Akad. Nauk SSSR, Ser. Fiz. 23(10):1184 (1959).

101. Ya. S. Bobovich and T. P. Tulub, Opt. i Spektroskopiya 6:566 (1959).

102. A. I. Stakhov and E. V. Chisler, Vestn. Leningr. Univ., Ser. Fiz. Khim. 4(1):159 (1959).

103. M. Harrand, Ann. Phys. 2(5-6):309-317 (1957).

104. W. B. Person, J. Chem. Phys. 23:2376 (1955).

105. V. M. Pivovarov, Opt. i Spektroskopiya 6:101 (1959).

106. V. M. Pivovarov, Opt. i Spektroskopiya 9:266 (1960).

107. B. S. Neporent and N. G. Bakhshiev, Opt. i Spektroskopiya 5:634 (1958).

108. J. Kirkwood, J. Chem. Phys. 4:592 (1936); 7:911 (1939); 8:205 (1940).

109. A. I. Ansel'm, Zh. Eksperim. i Teor. Fiz. 17:489 (1947).

110. M. V. Vol'kenshtein, Dokl. Akad. Nauk SSSR 32:185 (1941).

INVESTIGATION OF THE BROWNIAN ROTATIONAL
MOTION OF MOLECULES OF CONDENSED-PHASE SUBSTANCES
BY RAMAN SCATTERING AND INFRARED ABSORPTION METHODS

A. V. Rakov

INTRODUCTION

The study of the pattern of motion of molecules and of the nature of the intermolecular interaction in liquid and solid substances has become a major trend in contemporary molecular physics. Data on the interaction between molecules are required to secure information on the physical nature of the basic properties of matter, such as viscosity, compressibility, electrical conductivity, solubility, the kinetics of crystallization and polymerization, and similar properties. The investigation of the pattern of motion executed by molecules and the investigation of intermolecular interactions acquire added importance in the study of the liquid state of matter.

At the present time, statistical mechanics theory [1] is capable of accounting for the properties of only simple liquids. The practical application of this theory to complex liquids runs up against a number of serious obstacles, of which the most outstanding is the dearth of information on the intermolecular interaction potential and on the mechanism involved in the motion of molecules.

Current data on the nature of the liquid state indicate that the structure and physical properties of a real liquid are in powerful measure dependent upon the structure of the constituent particles of the liquid and on the nature of the forces acting between those particles. The sizes and shapes of the molecules and to an even greater degree the nature of the intermolecular interaction in turn decisively influence the nature of the relative position and relative motion of the particles of the liquid.

As a result, the problem of investigating the nature of molecular motion and of the effect of the intermolecular interaction of this motion comes to the fore in the study of the structure of liquids.

According to currently held concepts, the thermal motion of molecules in a liquid is represented in the form of comparatively infrequent transitions of the molecule from one equilibrium position to another equilibrium position and in the form of thermal vibrations or rotational oscillations between these transitions. The transition of a molecule from one equilibrium position to another may take place in the process as the consequence of a displacement of its center of gravity, or equally well as a consequence of a drastic change in the orientation of the molecule.

A variety of physical techniques are brought to bear in research on the mechanism underlying the motion of molecules in liquids and on the nature of the intermolecular interaction. The study of such properties as viscosity, solubility, electrical conductivity, adiabatic compressibility, and the like, enables us to secure some valuable information on the nature of the intermolecular interaction. At the present time, the method of nuclear magnetic resonance is being widely used in research on the mechanism underlying molecular motion in liquids.

---

* Dissertation in partial fulfillment of the requirements for the degree of Candidate in Physical and Mathematical Sciences. Defended before the Learned Council of the "Structure of Matter" panel of the Physical Faculty of the M. V. Lomonosov Moscow State University. Research Director: Doctor of Physical and Mathematical Sciences Professor P. A. Bazhulin.

Among the other techniques found popular in the study of the nature of the liquid state of matter, optical techniques have won themselves a prominent place. One of these is the method of Rayleigh scattering of light. M. A. Leontovich [2] has developed a relaxation theory which relates the intensity distribution in the wing of the Rayleigh line to the relaxation time of fluctuations in the anisotropy, which enables researchers to secure information on the motion of the molecules responsible for these fluctuations. Another optical technique which makes it possible to probe into the nature of the motion of molecules in liquids is provided by the polarized luminescence method. It has been established at this writing that this technique can be employed with success to obtain information on the rotational motion of molecules of the type known as Brownian motion.

A comparative new comer in the family of experimental research techniques described here is the study of the structure of liquids and of the nature of molecular motion based on the width of lines in the Raman spectrum. In 1953, L. L. Sobel'man [3] expressed a hypothesis on the effect of the random rotational motion of molecules in a liquid on the width of Raman lines. It was reported in that article that the amplitude of the scattered light wave will be modulated in response to random reorientations of the molecules if the derivative of the polarizability tensor with respect to the normal coordinate of that oscillation has an anisotropic part. This modulation of the amplitude of the scattered light wave leads to a broadening of the lines of the Raman spectrum, corresponding to the molecular vibrations in question. This broadening is in turn related to the mean lifetime of a molecule between two reorientations.

The developed theory of broadening of Raman lines, in the case of Raman scattering due to the Brownian rotational motion of the molecules, may be extended to the case of the broadening of infrared absorption spectral bands. Actually, the rotational motion of the molecules will result in random variations in the directions of the dipole moment of each molecular vibration. On the basis of the theory of random stationary processes, it may be shown that the absorption bands may be broadened in the process, but the amount of broadening involved will also be related to the mean lifetime of a molecule between two reorientations.

According to presently entertained notions on the thermal motion of molecules in liquids and in solids, the mean lifetime of a molecule between two reorientations is largely dependent upon the temperature and upon the intermolecular interaction. As a consequence, the study of the width of Raman lines and infrared absorption bands may be placed at the basis of a new method in research on the Brownian rotational motion of molecules.

In contrast to rival optical techniques, the use of Raman scattering of light and of infrared absorption for that purpose offers some striking advantages. First, the object of investigation in these instances is the molecule of the specific substance directly; second, these methods are sensitive only to changes in the orientation of a molecule, making it possible to study the role played by the "orientation jumps" of molecules in the general thermal motion; third, these techniques make it possible to study and compare the pattern of the rotational motion of molecules in liquids and solids, and fourth and last, these techniques can be utilized in the study of the rotational motion of molecules of the discrete components of various solutions. All of this opens up new opportunities for the study of the mechanism involved in molecular motion and for the study of the nature of the intermolecular interaction.

CHAPTER I

REVIEW OF THE LITERATURE AND STATEMENT
OF THE PROBLEM

The question of the relationship between the physical properties of liquids and solids and the mechanism underlying the thermal motion of the constituent molecules was first probed into in a detailed way by Ya. I. Frenkel' [4]. As the basis of his approach, he suggested the concept of a close affinity between the mechanism of thermal motion of molecules in liquids and crystals. Ya. I. Frenkel' pointed out the fact that molecules in the crystal lattice and in a liquid execute oscillations about their equilibrium positions, which exhibit a "temporal" character. The molecular transitions from one equilibrium position to another may occur either through the displacement of the center of mass of the molecule by a quantity of the order of magnitude of 10^{-8} cm, and as a result of a change in the orientation of the molecule. The fundamental characteristic of the mechanism of thermal motion is the mean lifetime of a molecule in the equilibrium position, τ. For example, in the case of a transition of a molecule from one equilibrium position to another due to a change in the molecule's orientation, the mean lifetime of the molecule in the equilibrium position will be

$$\tau_{or} = \tau' \exp\,(U_{or}/kT), \tag{I.1}$$

where τ' is the period of the "rocking" vibrations of the particle, k is the Boltzmann constant, T is the absolute temperature, and U_{or} is the potential barrier which a molecule has to climb over in executing a reorientation. The value of U_{or} is principally specified by the interaction of the specific molecule with its nearest neighbors. In what follows, the time τ_{or} will be known as the mean reorientation lifetime of the molecule, and will be denoted as τ_0.

Proceeding from this model as our point of departure, we are obliged to choose a method for our experimental investigation of the mechanism underlying the thermal random motion of molecules which will permit us to determine directly the mean lifetime τ of a molecule occupying an equilibrium position. The method of nuclear magnetic resonance will be suitable as one such method for determining τ.

It is well known that atomic nuclei possess an intrinsic angular momentum $\mathbf{I}$, known as the spin. These nuclei also possess the magnetic moments μ directed parallel (or antiparallel) to the mechanical moments. A simple relationship exists between the mechanical and magnetic moments, expressed by the equation:

$$\mu = \gamma \mathbf{I},$$

where γ is known as the gyromagnetic ratio. If the substance is immersed in a constant magnetic field H_0, then, as a result of interaction between the moments μ and the field H_0, the atomic nuclei will possess a system of equidistant energy levels with a spacing γH_0 between adjacent levels. If we switch on a variable electromagnetic field of frequency ω_0, an energy quantum $h\omega_0$ of that field will be capable of producing a jump between energy levels of the nucleus when the condition

$$\hbar\omega_0 = \gamma H_0 \tag{I.2}$$

is met. The frequencies ω_0 corresponding to these jumps will be known as resonance frequencies, and lie in the radio-frequency range.

The nuclear moments in matter interact with the thermal motion of the molecules (or atoms). This interaction has been given the name of lattice coupling in nuclear magnetic resonance theory. When an electromagnetic field is switched on, the energy associated with this field is absorbed as a result of the over-population of nuclei at the lower energy level. While the radio-frequency field acts to reduce the excess population of nuclei at the lower level, the coupling between the nuclear moments and the "lattice" tends to restore this excess to its equilibrium state. The approximation of the system of spins to the equilibrium state is known as a thermal relaxation process of characteristic time T_1, which has been dubbed the "spin-lattice relaxation time." This time is determined by experimental means.

In 1948, Bloembergen, Purcell, and Pound [5] laid the basis for, and experimentally verified, their theory of relaxation phenomena in nuclear magnetic resonance. They showed that the spin-lattice relaxation time T_1 is related to the mean lifetime τ of a molecule in its equilibrium position, and that for a large number of liquids τ coincides with the relaxation time, derived by Debye, for a Brownian spherical particle in a liquid:

$$\tau = \frac{4\pi a^3 \eta}{3kT} , \qquad (L3)$$

where a is the particle radius, k is the Boltzmann constant, T is the absolute temperature, and η is the viscosity of the liquid. From the findings of Bloembergen and other authors, we infer that the mechanism underlying the molecular motion in a liquid is an analog of the Brownian motion of particles, and that τ is related to the properties of a liquid — viscosity and temperature. At the present time, the nuclear magnetic resonance technique is being employed with success in research on the mechanism of thermal motion and on the structure of molecules, on the kinetics of chemical reactions, and a host of other major research problems [6].

Among the optical techniques enlisted in research on the mechanism behind molecular motion, the method of Rayleigh scattering of light has won great favor. As we know, the scattering line of liquids has monotonically decreasing wings, which sometimes extend to from 100 to 120 cm^{-1}. According to the theory developed by M. A. Leontovich [2], the distribution in the wing may be described by the formula

$$I(\Delta\omega) = \frac{C}{1 + (\Delta\omega)^2 \tau_p^2} , \qquad (L4)$$

where C is some constant, $\Delta\omega$ is the frequency computed from the center of the scattering line, and τ_r is the relaxation time of anisotropy fluctuations. We infer from Eq. (L4) that the function $1/I(\Delta\omega) = f(\Delta\omega^2)$ must be linear throughout the extent of the wing, but that τ may be determined from the slope of the line. It is natural to expect that the relaxation of anisotropy fluctuations will be associated with the thermal motions of the molecules.

In experimental terms, the study of the intensity distribution in the wing of the Rayleigh line has been carried out by I. L. Fabelinskii, E. F. Gross, M. F. Vuks, and others. I. L. Fabelinskii [7] made a painstaking study of the intensity distribution in the wing of the Rayleigh line, using interference techniques, and demonstrated that the entire extent of the wing does not yield to description by formula (L4). It was found that the function $1/I(\Delta\omega) = f(\Delta\omega^2)$ is linear over two segments, but shows a break in the curve in the frequency range of 8 to 10 cm^{-1} from the center of the line, i.e., there are two straight lines such that $1/I(\Delta\omega) = f(\Delta\omega^2)$, each of different slope, leading us to conclude that two relaxation times must exist. Measurement of the relaxation time of low-viscosity liquids in the region of the wing extending from 0 to 8 cm^{-1} revealed that this time is of the order of 10^{-12} to 10^{-11} sec (at room temperature), while measurements in the region of the wing from 8 to 60 cm^{-1} yielded a relaxation time of from 10^{-13} to $3 \cdot 10^{-13}$ sec, in order of magnitude. The viscosity dependence of the relaxation time was not investigated in that article. Highly viscous substances such as phenyl salicylate (salol), glycerin, and triacetin were studied. It was shown that the relaxation time increases as the viscosity, but at a slower rate than the latter. The author of that article draws the conclusion that the relaxation time of the anisotropy fluctuations does not coincide with the relaxation time of a Brownian spherical particle in a liquid.

Extensive research has been undertaken by E. F. Gross and M. F. Vuks [8, 9] on the wings of the Raman line. They expressed the hypothesis that the rocking vibrations of molecules affect a region of the wing far removed from the line center. In crystals, these vibrations appear in the form of the "near frequency" spectrum. In a liquid, the lines of this spectrum broaden to the point of overlapping, and form the far portion of the wing of the Rayleigh line. In subsequent contributions, M. F. Vuks and associates [10] developed the following viewpoint on the nature of the wing of the Rayleigh line: the region of the wing close to the line center is specified by the relaxation process described by M. A. Leontovich, while the remote portion represents a superposition on the intensity distribution of the "smeared-out" spectrum of molecular rocking vibrations due to the relaxation of the anisotropy fluctuations. The authors of those papers elaborated a method for determining the relaxation time of anisotropy fluctuations over the wing region adjacent to the center of the Rayleigh line. In their investigations, these authors proceeded from the assumption that the relaxation time τ_r coincides with the mean reorientation time of a molecule τ_0 in the liquid. The relaxation times they measured in low-viscosity liquids at room temperature lie in the range from 10^{-12} to 10^{-11} sec.

On the basis of these findings, we may conclude that the nature of the Rayleigh line wing is exceedingly intricate, and that the Rayleigh scattering method must be applied with care in the study of molecular motion in liquids.

Other optical techniques have also been used in studies of the rotatory motion of molecules in a liquid. Among these we may note the method of studying the Brownian rotatory motion of molecules by the polarization of the luminescent emission. This phenomenon first received detailed study in the investigations of S. I. Vavilov and V. L. Levshin [11]. Later on, V. L. Levshin went on to develop a theory of polarized luminescence [12] in which he treated the relationship linking polarization to the structure of the molecule and to the Brownian rotatory motion of molecules in solutions. According to this theory, the degree of polarization of the luminescent emission is related to the angle α through which the molecule rotates during the lifetime of an excited state, by the following formula:

$$\left(1 - \frac{3}{2}\sin^2\alpha\right)\left(\frac{1}{P} - \frac{1}{3}\right) = \left(\frac{1}{P_0} - \frac{1}{3}\right),$$ (I.5)

where P_0 is the limiting polarization characterizing the anisotropy of the molecule, and P is the polarization of the luminescent emission. In turn, the angle α is related to the characteristic properties of the medium (i.e., of the solvent) in which the molecule of the fluorescent substance is found. Taking this relationship into account, V. L. Levshin demonstrated that polarization of luminescence will be subject to the following law (known as the Levshin–Perrin formula):

$$\frac{1}{P} = \frac{1}{P_0} + \left(\frac{1}{P_0} - \frac{1}{3}\right)\frac{RT\tau_e}{V\eta},$$ (I.6)

where R is the gas constant, T is the absolute temperature, V is the volume of the molecule, τ_e is the lifetime of the excited state, and η is the viscosity coefficient of the solvent.

Formula (I.6) has been verified experimentally on more than one occasion [13, 14] for various substances; thus the linear relationship linking $1/P$ and T/η, and derived from the law (I.6), is well satisfied. These experiments permit us to infer that a molecule of a luminescent substance may be represented as a Brownian particle immersed in a liquid, and that the motion of this molecule yields to description by the laws of Brownian rotatory motion. The relationship between polarization of luminescent emission and the viscosity and temperature of the solvent, as revealed here, is one of the extent experimental proofs of the existence of some relationship between the physical properties of liquids and the mechanism underlying molecular motion.

New and broad possibilities for the investigation of the mechanism underlying the orientational motion of molecules in liquids by optical techniques has been opened up by a study by I. I. Sobel'man [3], in which the problem of the effect of the random rotatory motion of molecules, such as Brownian motion, on the width of Raman lines, was treated.

As is generally known, the intensity of Raman lines is proportional to the square of the amplitude of the scattered light wave. The amplitude of the scattered light wave E is in turn proportional to the derivative of the polarizability tensor α_{jk} with respect to the normal coordinate q_i corresponding to the given vibration of the molecule. If the tensor $\partial\alpha_{jk}/\partial q_i$ is an anisotropic tensor, then the amplitude E will be modulated by the Brownian rotatory motion of the molecules and will be a random function of time E(t). If the tensor $\partial\alpha_{jk}/\partial q_i$ is isotropic, reorientation of the molecule will be without effect on the value of E. In the general case, the tensor $\partial\alpha_{jk}/\partial q_i$ has both an isotropic and an anisotropic part. The relationship between these parts of the tensor is the determining factor in the dependence of the line profile on the Brownian rotatory motion of the molecules, i.e., the greater the anisotropy of the tensor $\partial\alpha_{jk}/\partial q_i$, the more pronounced will be the effect of random reorientations of the molecules on the line profile and on the line width. On the other hand, the integral depolarization ratio ρ of the Raman line is determined by this same relationship linking the isotropic and anisotropic parts of the tensor. This circumstance enables us to select the depolarization ratio of the Raman line as a distinctive criterion in judging the effect of the Brownian rotatory motion of molecules on the width of various spectral lines. Let us run through three possible cases.

1. The line is completely depolarized, $\rho = 6/7$; the trace of the tensor $\partial\alpha_{jk}/\partial q_i$ will vanish. The amplitude E of the scattered light wave is a random function of time characterized by the correlation function [15]:

$$\varphi(\tau) = \overline{E(t)E^*(t+\tau)}. \tag{I.7}$$

The function $\varphi(\tau)$ attains a maximum at $\tau = 0$, and declines monotonically as τ increases. As $\tau \to \infty$, the function $\varphi(\tau) \to 0$. For all practical purposes the correlation vanishes at some $\tau = \tau_k$. The customary assumption in dealing with such processes as the process of random molecular reorientations in the theory is that:

$$\varphi(\tau) = \overline{E^2}e^{-\frac{|\tau|}{\tau_k}}. \tag{I.7a}$$

According to the theory of stationary random processes, the correlation function is related to the spectral density of the process, i.e., in the specific case it is related to the spectral intensity distribution, by the following formula [15]:

$$I(\omega - \omega_0) = \frac{1}{\pi}\,\mathrm{Re}\int_{-\infty}^{\infty}\varphi(\tau)e^{-i\,(\omega-\omega_0)\,\tau}\,d\tau. \tag{I.8}$$

From the solution of (I.8) with the substitution of the value of $\varphi(\tau)$ from (I.7a), we find the width of the Raman scattering line to be*

$$\Delta\omega = 2/\tau_k. \tag{I.9}$$

Or, in the transition from the cyclic frequency to wave numbers,

$$\Delta\nu = 1/\pi c\tau_k, \tag{I.10}$$

where c is the speed of light. Since the correlation function in this case characterizes the random process of molecular reorientation, the time τ_k will be the mean lifetime of a molecule between two successive reorientations or, as we agreed to term it, the mean reorientation time of the molecule τ_0.

Strictly speaking, there may exist several such times τ_{0i} for a specific molecule, these times being due to the fact that there is a different molecular reorientation probability about the several axes of rotation available to the molecule. For molecules having specified directions of rotation (such as long paraffin chains), then, the main contribution to line broadening would be that contributed by flips of the molecules about their most probably axes of rotation. In the case of molecules exhibiting no specific discernible axes of rotation (i.e., symmetric molecules), the times τ_{0i} will be the same no matter what axis serves for the reorientation.

2. $0 < \rho < 6/7$; the tensor $\partial\alpha_{jk}/\partial q_i$ may be broken up into isotropic and anisotropic components. Accordingly, an anisotropic part of width $\Delta\nu = 1/\pi c\tau_k$ may be discerned in the line, and an isotropic part whose width is independent of the Brownian rotatory motion of the molecules is also present.

---

*Here we abstract from the effect of other factors on broadening of lines in the Raman spectrum.

3. $\rho = 0$; the tensor $\partial \alpha_{jk} / \partial q_i$ is isotropic and Brownian rotatory motion of the molecules will have no effect on the width of the lines.

The jump of a molecule from one equilibrium position to another in the form of a shift of its center of mass with no change in orientation will not modulate the amplitude of the scattered light wave for all oscillations of the molecule. Doppler broadening due to diffusion of the molecules (phase modulation) is so slight in the case of interest to us that it may be safely disregarded.

On the basis of the foregoing discussion, we may expect that the greatest changes in line width in response to any variation in the external conditions in which the substance is found will be observed in the depolarized Raman lines. Measurements of the width of the depolarized Raman lines opens up the possibility of using (I.10) in order to determine an important parameter of the Brownian rotatory motion of molecules: the mean reorientation time τ_0 of the molecule.

The arguments advanced by I. I. Sobel'man on broadening of Raman spectral lines due to the Brownian rotatory motion of molecules may be extended to the case of broadening of infrared absorption bands. In actuality, the integral absorption coefficient for a given vibration of the molecule is a function of the value of the projection of the derivative with respect to the normal coordinate of the dipole moment onto the axes of the fixed coordinates of observation [16]. As the molecule assumes a new orientation the direction of the dipole moment of that mode of vibration is altered. As a result, there will also be changes in the value of the projections of the derivative of the dipole moment onto the coordinate axes of observation. As a consequence of that in turn, the value of the projections of the derivative of the dipole moment due to random flips of the molecules will be a random time function. Taking into consideration the known relationship between the spectrum of the random function and its correlation function, we may show that the width of the spectral distribution of the integral absorption coefficient, due to the Brownian rotatory motion of the molecules, will be [15]:

$$\Delta\omega = 2/\tau_0, \tag{I.11}$$

where τ_0 is the mean reorientation time of the molecules. As in the Raman scattering case, therefore, the infrared absorption bands will be broadened on account of the random reorientations by the amount

$$\Delta\nu = \frac{1}{\pi c \tau_0} \text{ cm}^{-1}. \tag{I.12}$$

The relationship linking the width of the depolarized Raman lines and infrared absorption bands and the fundamental characteristic τ_0 of the Brownian rotatory motion of the molecules opens up a whole range of new opportunities for the study of the orientational motion of molecules, using the Raman scattering and infrared absorption techniques of spectroscopy. These methods enable experimental researchers to determine the mean reorientation time of the molecule in an experiment where the molecular species is directly the object of investigation, and to study the dependence of that time on the temperature and viscosity of the substance. The investigation of the width of Raman lines and infrared absorption bands as the substance is converted from the liquid state to the crystalline and vitreous states, in the transition from homogeneous liquids to solutions, and so forth, may yield much valuable information on the effect of intermolecular coupling on the nature of the orientational motion of molecules. A comparison of the mean reorientation time of the molecule and the relaxation time of anisotropy fluctuations determined over the near region of the wing of the Rayleigh exciting line may yield additional information on the physical nature of the wing.

At the time work was begun on this thesis (1953), there were no experimental contributions to the literature confirming the theoretical points advanced by I. I. Sobel'man. Worse yet, there had been no attempts made to apply the techniques of Raman scattering of light and infrared absorption to research on the mechanism underlying the rotatory Brownian motion of molecules. Realizing full well that we are only scratching the surface in research to date on the orientational motion of molecules, and on reviewing the possible new approaches now opening up in the study of this problem, we decided to the pose the following questions in our work.

1. Experimental verification of the possibility of applying Raman and infrared absorption techniques in research on the Brownian rotatory motion of molecules of substances in the condensed state. This requires a study of the dependence of the width of Raman lines of varying depolarization ratio and of the width of absorption lines on the temperature and state of aggregation of the substance.

2. Elaboration of a method for determining the mean reorientation time τ_0 of the molecule.

3. Elucidation of the effect of the intermolecular interaction on the Brownian rotatory motion of molecules, in order to investigate the relationship between Raman widths and infrared absorption bandwidths, on the other hand, and the viscosity of liquids and the liquid—solid phase transition on the other.

4. Elucidation of the nature of molecular reorientations in crystalline substances.

5. Performing an analysis of the resulting findings, drawing upon data available in the literature on the Brownian motion of molecules as secured by other techniques.

To this purpose, studies were made of the Raman spectra and infrared absorption spectra of various organic compounds of a variety of paraffins, aromatic and cyclic compounds, and alcohols. The spectra were studied at various temperatures in the liquid and solid states.

CHAPTER II

EXPERIMENTAL INVESTIGATION
OF THE BROWNIAN ROTATORY MOTION
OF MOLECULES IN LIQUIDS
BY MEANS OF THE RAMAN SCATTERING METHOD

§1. Equipment and Techniques for Measuring Raman Line Widths

The Raman spectra were recorded by both photographic and photoelectric means. A three-prism HUET B-III spectrograph was employed in the photograph. The dispersing part of the instrument consists of three 60° prisms fabricated of heavy flint. The prisms stood 90 mm high, on a 170 mm base. The focal length of the collimator objective was 900 mm; the focal length of the camera objective was 650 mm. The diameter of the collimator and camera objectives was 100 mm. The linear dispersion of the instrument in the λ = 4358 A region was 8 A/mm, and was 6 A/mm in the λ = 4047 A region. The linear magnification of the spectrograph, as measured experimentally, was 0.7 in the λ = 4358 A region, and 0.9 in the λ = 4047 A region. The instrument housing was double-walled to thermostat the dispersing system. In order to keep the temperature of the prisms constant, the water was constantly circulated from a TC-15 ultrathermostat kept at a present temperature inside the spectrograph housing, during the entire course of the experiment.

Low-pressure mercury lamps with water-cooled liquid electrodes were employed as excitation sources for the Raman spectra. Two designs of these lamps were utilized in our work: one design developed by M. L. Sosinskii [17], another, with a double luminous channel, described in an article by N. L. Rezaev [18].

In order to measure the true width of Raman lines, we have to take into consideration the distortions introduced in the recorded intensity distribution of a finite width of the Rayleigh excitation line by the instrument and by the photographic plate. These distortions are described by the apparatus function. The relationship linking the true intensity distribution $\varphi(y)$ in the line and the observed intensity distribution F(x) is given by the convolution equation

$$F(x) = \int_{-\infty}^{\infty} \varphi(y)\, a\,(x-y)\, dy,$$ (II.1)

where a(x — y) is the apparatus function. When the true line width is measured, a(x — y) must be known. The apparatus function was determined by experimental means in the work described here.*

The distortions in the observed line profile as introduced by the finite dimensions of the spectrograph slit were eliminated by the Rayleigh graphical method [19]. This method is described in the literature [20],

---

* The investigation of the equipment function applying to photographic recording was carried out jointly with G. V. Mikhailov (cf. the dissertion of G. V. Mikhailov in this volume).

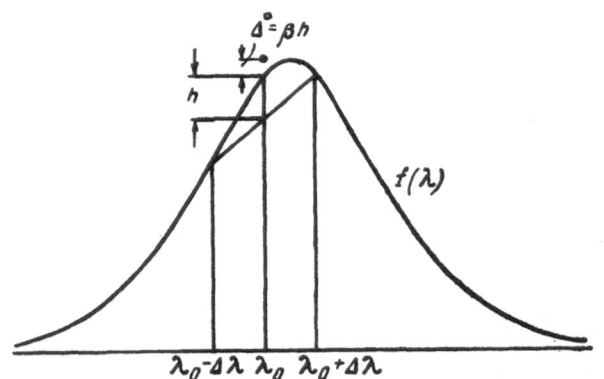

Fig. 1. Diagram illustrating the treatment of the apparatus slit function in the Rayleigh method.

and we shall not go into it in any further detail. Here we present only a general set of rules governing the use of this method to eliminate distortions in the line contour.

Let $f(\lambda)$ be the observed line profile (Fig. 1). In order to take the distortions introduced by the finite slit width of the instrument into account (i.e., to take the apparatus slit function into account), we mark off a sequence of segments $\pm \Delta\lambda$ from each point λ_0. A straight line is drawn through the points $f(\lambda_0 - \Delta\lambda)$ and $f(\lambda_0 + \Delta\lambda)$, cutting off a segment h on the ordinate $\lambda = \lambda_0$. The correction introduced into the line profile will be $\Delta^0 = \beta h$, where β is a coefficient dependent upon the form of the equipment function and on the choice of interval $\Delta\lambda$. The segment $\Delta^0 = \beta h$ is always laid off in the direction of the convex side of the curve $f(\lambda)$. The coefficient β is determined in the following manner (where α is the width of the apparatus function):

1) when the apparatus function is of rectangular form (spectrograph slit):

$$\beta = \frac{1}{12} \frac{\alpha^2}{(\Delta\lambda)^2} ;$$

2) when the apparatus function is of triangular shape (two monochromator slits):

$$\beta = \frac{1}{6} \frac{\alpha^2}{(\Delta\lambda)^2} ;$$

3) when the apparatus function is of Gaussian shape:

$$\beta = \frac{1}{5.5} \frac{\alpha^2}{(\Delta\lambda)^2} .$$

Consequently, we have $\alpha = S$ when the spectrograph is of slit width S; choosing $\Delta\lambda = S/2$, we have $\Delta^0 = \frac{1}{3}h$ for our correction. In the case of the monochromator, $\Delta\lambda$ is taken equal to $s/\sqrt{2}$, so that the correction will again be $\Delta^0 = h/3$.

An analytical function describing the observed profile corrected by the Rayleigh method, and an analytical function describing the measured apparatus function, must be substituted, in the general case, in Eq. (II.1), in order to determine the true Raman line width. Equation (II.1) was solved for several special cases. Figure 2 presents a graph of the convolution of the true line profile of dispersion form with the profile of the apparatus function exhibiting a dispersion form (straight line 1) and a Gaussian form (curve 2). By availing ourselves of this graph, we shall be in a position to determine the true line width from the width of the observed profile and from the width of the apparatus function.

An analysis of the observed Raman profiles revealed that their shape is close to the dispersion profile, i.e., the convolution of the dispersion line with the dispersion equipment function may be used, thereby yielding

$$\delta_{true} = (\delta_{obs} - \delta'_{eq})D \text{ cm}^{-1}, \tag{II.2}$$

where δ_{obs} is the width of the observed profile of the line under investigation, corrected by the Rayleigh method to eliminate the distortions introduced by the spectrograph slit (in mm of the plate); δ'_{eq} is the width of the apparatus function (in mm length of plate) and D is the linear dispersion of the instrument for the specific wavelength (cm^{-1}/mm). The deviation of the profile of the equipment function from the dispersion shape

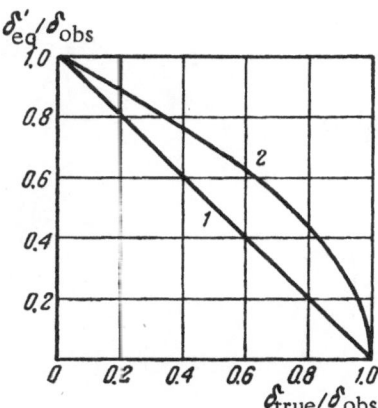

Fig. 2. The graph of the convolution of the profile of a line of dispersion shape with the profile of the apparatus function. 1) Apparatus function of dispersion shape; 2) apparatus function of Gaussian shape. δ'_{eq}) Width of apparatus function; δ_{obs}) width of observed profile; δ_{true}) width of true profile.

leads to a correction which lies within the range of error of the measurements, and will be ignored in the subsequent discussion. The total average error in the determination of the true line width, including the error in the measurements and the error incurred in computing by formula (II.2), comes to ±0.5 cm^{-1} in the case of narrow lines, and to ±0.3 cm^{-1} in the case of broad lines.

The Raman spectra were recorded photoelectrically on a DFS-12 spectrometer. The spectrometer was a double monochromator with diffraction grids. The optical layout of the instrument may be seen in Fig. 3. The principal specifications of the spectrometer are the following: operating range 3600 to 6500 A; focal length of parabolic reflector objectives 800 mm; relative to collimator aperture 1:5.3; the measured linear dispersion over the exit slit 4.6 A/mm (constant for all wavelengths); diffraction grids (replicas) were flat with 600 lines/mm. The radiation sensor used here was a FÉU-17 photomultiplier tube. The signal from the photomultiplier was fed to a DC amplifier. The spectra were recorded on a PS1-02 strip-chart recording potentiometer.

The apparatus function of the instrument was studied on the lines of the mercury spectrum, the source for which was a low-pressure mercury tube. The form and width of the apparatus function was measured by the line λ = 5769 A, which has what is practically a symmetrical profile. When recording the line, the entrance and exit slits of the spectrometer are set at 3 μ (or the order of a normal slit width). The results of measurements have shown that the shape of the profile of the apparatus function is very close to a Gaussian shape, and its width is approximately 0.3 cm^{-1}. Figure 4 shows a spectrogram of the trace of the profile of that line. Calculations of the width of the apparatus function at other wavelengths, with allowance for instrument dispersion, showed that the width of the apparatus function must be δ'_{eq} = 0.35 cm^{-1} for λ = 5461 A, and δ'_{eq} = 0.7 cm^{-1} for λ = 4358 A.

Profiles of the line λ = 4916 A were investigated at various widths of the entrance and exit slits of the spectrometer in order to study the slit apparatus function. The center transmission slit was set at three to four times wider than the entrance slit, in order to eliminate any possible errors due to inaccurate setting of the double monochromatization. Measurements showed that the slit apparatus function of the double monochromator is twice as narrow as the slit apparatus function of a single monochromator with slits set at different widths. This result is found to be in complete agreement with theory. It is not difficult to show that the width of the apparatus function in the case of a single monochromator will be, in millimeters of paper:

$$L_1 = \frac{Sv}{\left(\frac{d\varphi}{dt}\right)F},$$ (II.3)

where S is the geometric width of the entrance or exit slit, v is the rate of feed of the recording coordinate paper, dφ/dt is the angular velocity at which the grid is rotated, and F is the focal length of the collimator objective. In the case of a double monochromator, the width of the slit apparatus function, in millimeters of paper, will be

$$L_2 = \frac{Sv}{2\left(\frac{d\varphi}{dt}\right)F}.$$ (II.4)

Hence, $L_1/L_2 = 2$.

In all further measurements, we made use of the total apparatus function (slit apparatus function, non-slit apparatus function, and profile of the exciting line), which was determined experimentally. Figure 5 shows

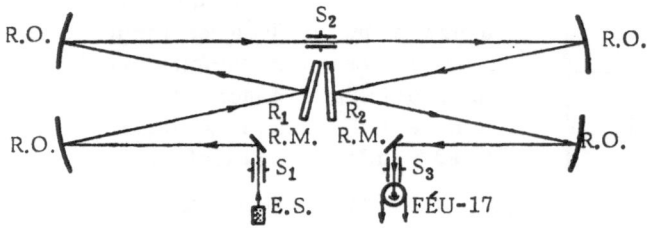

Fig. 3. Optical layout of DFS-12 spectrometer. S_1, S_2, S_3) Slits; R.O.) reflector objectives; R_1, R_2) replicas; R.M.) rotating mirrors; E.S.) excitation source.

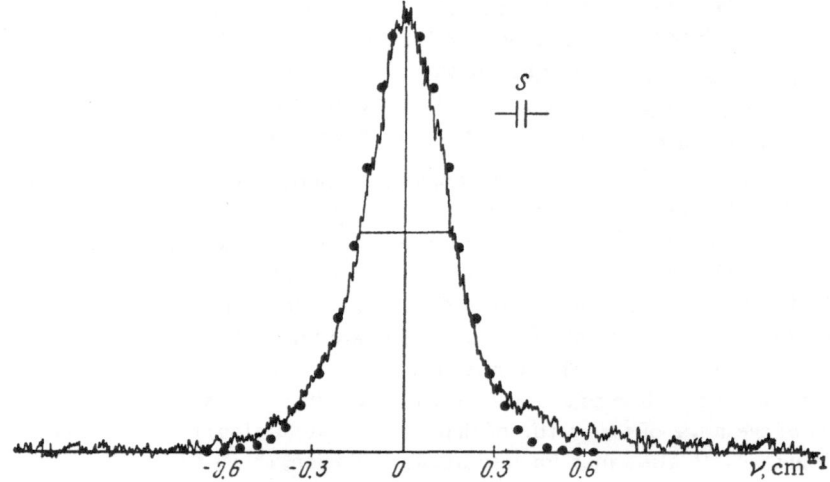

Fig. 4. Nonslit apparatus function of the DFS-12 spectrometer (points indicate the Gaussian curve).

TABLE 1. Comparison of Results of Measurements of Raman Line Widths in the Benzene Spectrum, by Photoelectric and Photographic Methods.

Line frequency, cm^{-1}	Line width, cm^{-1}			
	Photographic method		Photoelectric method	
	20°C	50°C	20°C	50°C
606	6.6	7.8	6.9	7.5
992	1.7	1.9	1.8	2.0
1586	15.0	15.0	12.5	15.8

a graph of the dependence of the width of the total apparatus function δ_{eq} on the geometric width of the slits $S_1 = S_3 = S$ for different scanning speeds. The shape of the profile of the total apparatus function proved to be close to the Gaussian shape in the case of our instrument.

Scanning speeds of 0.241, 0.462, and 0.885 A/min were used in recording the Raman lines. Measurement of the true width of the Raman lines was carried out as follows. The width of the observed profile of

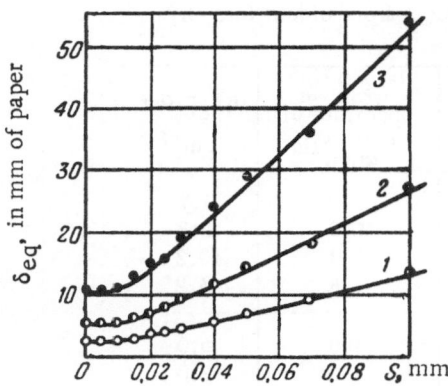

Fig. 5. Dependence of width of apparatus function on width of spectrometer slits $S_1 = S_3 = S$ at various scanning speeds. 1) 0.462 A/min; 2) 0.241 A/min; 3) 0.124 A/min; paper feed rate 720 mm/h.

the line was measured. According to the graph (Fig. 5), the width of the total apparatus function was determined for selected spectrometer slit widths and a specified scanning speed. Bearing in mind the fact that the Raman lines display a dispersion form, while the total apparatus function displays a Gaussian form, we found the true width of the line on the graph of the convolution of the dispersion profile with the Gaussian profile (cf. Fig. 2, curve 2).

Table 1 below, in which Raman data for the benzene spectrum are listed at 20° and 50°C, is provided to facilitate a comparison of the results of line width measurements by the photoelectric method and by the photographic method.

It is clearly discerned from this table that the results secured by the two methods fit very well.

In all of our subsequent researches, we made use of both the photoelectric and photographic methods in recording our spectra.

§2. Investigation of the Dependence of Line Width on Aggregate State, Temperature, and Viscosity of the Substance

The effect of the Brownian rotatory molecular motion on the width of Raman lines may be discovered experimentally by studying the dependence of the line width on the aggregate state, temperature, and viscosity of the substance. Actually, as the temperature of the liquid increases (or as the viscosity diminishes), the probability of reorientation on the part of the molecule will increase substantially, and the rotations of the molecules will be either strongly hindered or else will change their character in the course of the transition from a liquid to a solid, whereupon a change in the width of the Raman lines must result. The first investigations into the effect of the rotatory motion of molecules on Raman line width were carried out by the authors studying the spectra of benzene and paradichlorobenzene. In our work we measured the width of Raman lines of differing depolarization ratios as the substance underwent its transition from the liquid to the crystalline phase.

The spectrum of benzene was taken at 18°C in the liquid case and at −5°C in the crystalline phase. The melting point of benzene is 5.6°C; the melting point of paradichlorobenzene is 52.7°C; its liquid-phase spectrum was taken at 54°, the crystalline at 52°. Table 2 shows the results of the measurements. As we clearly see from the table, the line width of the spectra investigated in the transition from the liquid to the crystalline phase changes stepwise, and the greatest change in width is observed in the case of lines of depolarization ratio $\rho \approx ^6/_7$. The width of lines of low depolarization ratio ($\rho \approx 0.06$) changes only imperceptibly.

Similar results were later reported by N. I. Rezaev [21], who studied the liquid−crystal transition in the case of para-xylene and cyclohexane.

The change in line width at different polarizations in the transition of a substance from the liquid to a vitreous state has been studied by the authors in investigations of the spectra of ethyl alcohol, 2-methylpentane, and 2-methylhexane [22]. These substances vitrefy at low temperatures T_g, which are −114°, −155°, and −118°C respectively. The line width of the Raman spectra of these substances was measured over a broad range of temperatures. Graphs of the temperature vs. Raman line width are presented in Figs. 6 to 8. Clearly, from inspection of the graphs, we see that the Raman line widths of these substances decrease as the temperature of the respective liquid is lowered; the greatest change in line width is observed in the case of lines of depolarization ratio $\rho \approx ^6/_7$. The width of Raman lines have low ρ values varies only imperceptibly in response to a change in the temperature of the liquid. As the substance passes from the liquid to the vitreous state, no abrupt changes of any kind are observed in the line width, no matter what the value of ρ. In the range of temperatures close to the vitrefying point T_g, the width of all the Raman lines either remains prac-

TABLE 2. Change in Width of Raman Line as Substance Goes from Liquid into Crystalline Phase

Substance	Line frequency cm^{-1}	Depolarization ratio ρ	Raman line width in spectrum of liquid σ_{liq}, cm^{-1}	Raman line width in spectrum of liquid σ_{liq}, cm^{-1}	$\sigma_{liq} - \sigma_{cr}$, cm^{-1}
Benzene	606	0.86	6.6	1.9	4.7
	992	0.06	1.7	0.9	0.8
	1586	0.81	12.5	2.3	10.2
	3062	0.30	7.8	3.6	4.2
Paradichloro-benzene	298	0.73	6.0	1.0	5.0
	331	0.21	3.0	1.0	2.0
	627	0.86	2.5	1.0	1.5
	747	0.06	2.5	2.0	0.5
	1578	0.67	5.0	4.0	1.0
	3064	0.86	11.0	6.0	5.0
	3072	0.39	8.5	5.0	3.5

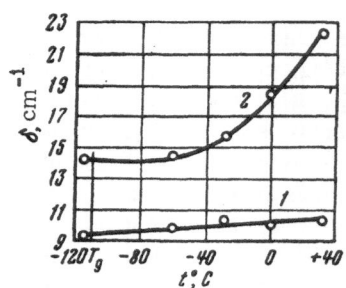

Fig. 6. Temperature vs. line width, spectrum of ethyl alcohol. 1) ν = 880 cm^{-1}, ρ = 0.22; 2) ν = 1040 cm^{-1}, ρ = 0.52; T_g = −114°C.

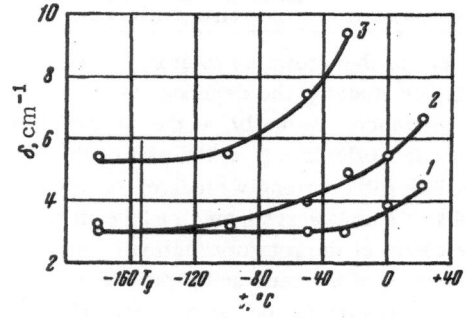

Fig. 7. Temperature vs. line width, spectrum of 2-methylpentane. 1) ν = 815 cm^{-1}, ρ = 0.2; 2) ν = 1039 cm^{-1}, ρ = 0.4; 3) ν = 1449 cm^{-1}, ρ = 0.85; T_g = −155°C.

tically constant or else undergoes only a very slight change. The width of different lines still remains very appreciable (from 3 to 15 cm^{-1}). There are apparently still a few more reasons for broadening of the lines investigated which may have nothing in particular to do with the Brownian rotatory motion of molecules in the liquid. The physical nature of these causal factors remains obscure at this writing. In what follows, this will be termed the "residual width."

Similar regularities were discovered subsequently by other authors [23] in the spectra of liquids vitrefying at a low temperature.

The relationship between the width of various Raman spectral lines and the temperature of the liquid were also studied by the authors in the case of benzene and toluene spectra. As an example, Fig. 9 presents a graph of the temperature dependence of the Raman width of toluene. The reader will clearly discern from the graph that the greatest change in width in response to a temperature variation is observed in the case of depolarized Raman lines.

The results reported of our measurements and the measurements by other authors [18, 21, 23, 24] reveal that as the temperature drops and as the substance transforms from the liquid to the crystalline phase, an appreciable decrease is observed in the width of depolarized Raman lines ($\rho \approx {}^6\!/_7$), while the width of lines

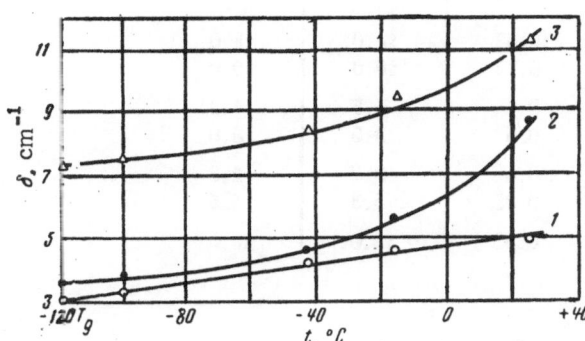

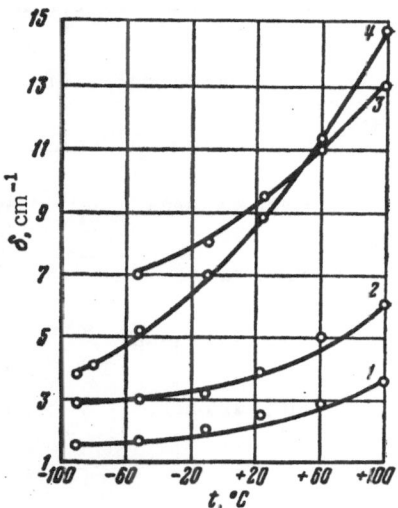

Fig. 8. Temperature vs. line width. Raman spectrum of 2-methylhexane. 1) ν = 849 cm^{-1}, ρ = 0.4; 2) ν = 1144 cm^{-1}, ρ = 0.6; 3) ν = 1304 cm^{-1}, = 0.74, T_g = −118°C.

Fig. 9. Temperature vs. line width, Raman spectrum of toluene. 1) ν = 786 cm^{-1}, ρ = 0.09; 2) ν = 1211 cm^{-1}, ρ = 0.13; 3) ν = 217 cm^{-1}, ρ = 0.87; 4) ν = 623 cm^{-1}, ρ = 0.73.

having a low depolarization ratio varies only insignificantly; the liquid—crystal transition is accompanied by a stepwise change in line width, we may note. Furthermore, as experimentally derived data reveal, the profiles of the depolarized Raman lines display a form close to the dispersion form. All of these experimental results turn out to be in excellent concordance with the inferences drawn in a paper by I. I. Sobel'man [3], and are indicative of some relationship with the Brownian rotatory motion of molecules.

The reported experimental data indicate a possible relationship between the width of depolarized Raman lines and the viscosity as well, since the viscosity of liquids is largely dependent upon the mobility of the constituent molecules of the liquid [4]. In the case of Brownian rotatory molecular motion, the relationship between the relaxation time of a spherical particle and the viscosity is quite well known in the form expressed by Debye's formula [cf. (L3)]. This formula is invoked in many problems in physics to evaluate molecular characteristics. Highly satisfactory findings have resulted from its use. If we suppose that this relationship is indeed valid in our case as well, then we may expect the mean reorientation time τ_0 to be proportional to η/T. Allowing for the fact that the viscosity decreases at a far more rapid rate than the temperature, we should expect, in actual practice, some such relationship as $\tau_0 \approx \eta$. Since the time τ_0 is related to the width of the depolarized Raman lines by the formula (I.12), the width of the lines must be proportional to the reciprocal of the viscosity coefficients.

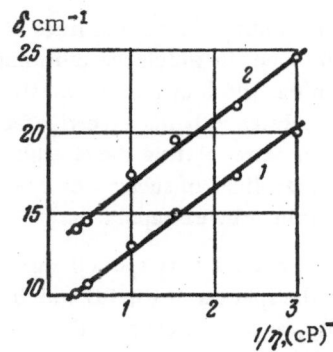

Fig. 10. Dependence of Raman line width of cyclopentane spectrum on reciprocal viscosity $1/\eta$. 1) ν = 1449 cm^{-1}, ρ = 0.72; 2) ν = 1031 cm^{-1}, ρ = 0.82.

Figures 10 to 12 provide us with examples of graphs plotting the width of depolarized Raman lines as a function of the reciprocal viscosity $1/\eta$ in the case of the spectra of cyclopentane, isopentane, and metaxylene. The viscosity values were taken from handbooks [25, 26]. As we see on the graphs, the line width is a linear function of the reciprocal viscosity. The slope of the straight-line plots $\delta = f(1/\eta)$ for various lines of the same Raman spectrum depends on the depolarization ratio ρ of the line studied. As a rule, the greater the value of ρ, the steeper will be the slope.

TABLE 3. Width of Raman Lines in the Raman Spectra of Vitreous Substances

Substance	Raman frequency, cm^{-1}	Depolarization ratio, ρ	Line width in spectrum of glass, cm^{-1}	
			direct measurements	by extrapolation
Cyclopentane	1031	0.82	14.0	13.0
	1449	0.72	10.0	9.0
Isopentane	909	0.7	4.5	4.3
	954	0.5	3.5	3.0
2-methylpentane	1039	0.4	3.2	2.3
	1440	0.85	5.3	4.5
Ethyl alcohol	1040	0.52	14.0	13.5

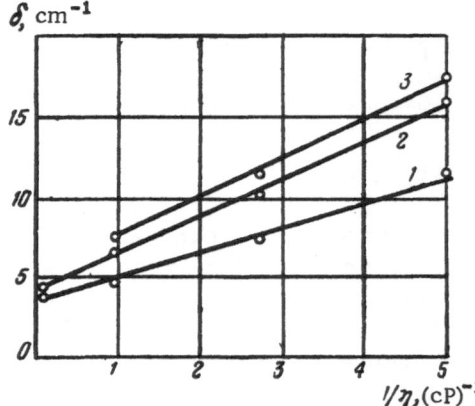

Fig. 11. Raman line width as a function of $1/\eta$ in the isopentane spectrum. 1) $\nu = 954$ cm^{-1}, $\rho = 0.5$; 2) $\nu = 909$ cm^{-1}, $\rho = 0.7$; 3) $\nu = 1147$ cm^{-1}, $\rho = 0.7$.

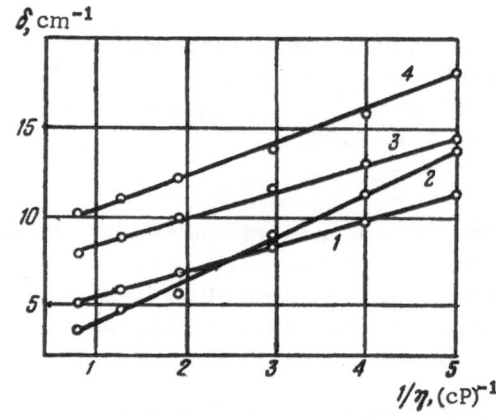

Fig. 12. Raman line width as a function of $1/\eta$ in the metaxylene spectrum. 1) $\nu = 1379$ cm^{-1}, $\rho = 0.59$; 2) $\nu = 1171$ cm^{-1}, $\rho = 0.84$; 3) $\nu = 517$ cm^{-1}, $\rho = 0.60$; 4) $\nu = 279$ cm^{-1}, $\rho = 0.65$.

The experimental results obtained demonstrate that there exists such a relationship $\tau_0 \approx \eta$ referable to the mean reorientation time τ_0, just as in the case of the Debye formula.

Extrapolating the lines in Figs. 10 to 12 out to the value $1/\eta$ corresponds to finding the Raman lines of the liquid at infinitely high viscosity, i.e., the case of a vitreous state of the substance, in which the Brownian rotatory molecular motion will have virtually ceased to exert any more effect on the width of the Raman lines. We should therefore anticipate that the widths of the Raman spectral lines of the substance being investigated in the vitreous state will agree with the width of those same lines as obtained from extrapolating the straight-line plots $\delta = f(1/\eta)$ out to $1/\eta = 0$. In Table 3, we have the widths of the Raman lines of substances occurring in the vitreous state and the corresponding values of the line widths as obtained by extrapolation.

The method described here, involving extrapolation of the lines $\delta = f(1/\eta)$, bears a very special significance in attempts to find the "residual width" of depolarized Raman lines in the spectra of liquids which crystallize in their transition to the solid phase. Such liquids possess low viscosity, as a rule, even in the vicinity of their melting points. For example, benzene exhibits a viscosity 0.008 P near its melting point, paraxylene a viscosity 0.007 P, metaxylene 0.015 P, etc. At such low viscosities, the role of the Brownian rotatory movement of molecules must be appreciable in the broadening of the Raman lines; the width of the lines measured in the Raman spectrum near the respective melting points could not be assumed, on that ac-

126

count, to be equal to the "residual width." Since the nature of the "residual width" remains quite obscure at this point, there are generally speaking no grounds supporting the assumption that this width will not change in the transition from a liquid to a crystal, i.e., the determination of the "residual width" is not unique in the case of the Raman spectrum of a liquid with respect to the spectrum of the corresponding crystal. The method of extrapolation described here appears to be the most convenient and most reliable approach to deal with this contingency.

The experimental data presented in this section demonstrate the fact that the profiles of the depolarized Raman lines exhibit a form close to the dispersion shape, whereas their width depends as a rule on the depolarization ratio ρ. When the temperature and viscosity of the liquids are measured, the Raman widths are discovered to vary most drastically when $\rho \approx {}^{6}/_{7}$. It has been found that the relationship linking the mean reorientation time τ_0 of the molecule to the viscosity is well described by the Debye formula. In the transition of the substance from the liquid state to the crystalline state, we observe a stepwise change in line width, and in the transition from the liquid state to the vitreous state we observe, in contrast, a smooth gradation of the width. All these data are indicative of some connection between the phenomena observed and the Brownian rotatory motion of the molecules, and this inference is in excellent accord with the conclusions arrived at by Sobel'man [3].

§3. Determination of the Mean Reorientation Time and of the Reorientation Potential Barriers of the Constituent Molecules in Liquids

Our experimental researches have revealed that the width of Raman spectral lines consists of two parts: a first temperature-dependent part related to the Brownian rotatory motion of the molecules in some way, and a second part which is constant and independent of the motion executed by the molecules (this part is known as the "residual width"). The total width of the depolarized Raman lines may then be written in the following manner*:

$$\delta_i(T) = \delta_{0i} + \Delta\nu(T), \tag{II.5}$$

where $\delta_i(T)$ is the width of the Raman line at a specified temperature T of the liquid, $\Delta\nu(T)$ is the width of the line or that part of the width due to the Brownian rotatory movement of the molecules, and δ_{0i} is the "residual width." Since experience has shown the "residual width" to differ for different lines in the Raman spectrum, we shall indicate the subscript "i" in the further discussion for the width of the line corresponding to the i-th vibration.

Substituting the value $\Delta\nu(T) = 1/\pi c\tau_0(T)$ from formula (I.12) into Eq. (II.5), we have

$$\tau_0 = \frac{1}{\pi c [\delta_i(T) - \delta_{0i}]}. \tag{II.6}$$

Consequently, in order to calculate the mean reorientation time τ_0 of the molecule, we are obliged to measure the width of the depolarized Raman line and to determine the "residual width" δ_{0i} for that line. The way to determine δ_{0i} was described in the preceding section.

In Table 4 we have the mean reorientation times τ_0 for several liquids, at 20°C, from formula (II.6). Clearly, from the table, the mean reorientation times of a molecule in the liquid will be of the order of magnitude of 10^{-12} sec, in the case of the substances investigated. Moreover, as the structure of the molecule, in substances belonging to the same homologous series, becomes more complex, the reorientation time τ_0 will be observed to increase systematically. For example, $\tau_0 = 2.0 \cdot 10^{-12}$ sec in the case of benzene, $\tau_0 = 3.0 \cdot 10^{-12}$ sec in the case of toluene, and $\tau_0 = 4.0 \cdot 10^{-12}$ sec in the case of paraxylene. A similar regularity is also observed in the case of substances with membership in the paraffin series: $\tau_0 = 1.1 \cdot 10^{-12}$ sec in the case of isopentane, $\tau_0 = 1.2 \cdot 10^{-12}$ sec in the case of 2-methylpentane, and $\tau_0 = 2.5 \cdot 10^{-12}$ sec in the case of 2-methylhexane.

---

*Formula (II.5) is valid whenever the profiles of lines of width δ_{0i} exhibit a dispersion shape. Specially designed low-temperature experiments have demonstrated that these profiles are adequately described in practice by a dispersion curve.

TABLE 4. Mean Reorientation Time of Molecule in Liquid for Several
Substances at 20°C

Substance	Raman frequency, cm^{-1}	Depolarization ratio, ρ	Mean orientation time of molecule, $\tau_0 \cdot 10^{12}$, sec
Benzene	606	0.88	2.0
	1486	0.81	
Toluene	623	0.73	3.0
	1605	0.70	
Metaxylene	517	0.60	4.3
Paraxylene	318	0.94	4.0
	645	0.89	
	1616	0.72	
Isopentane	909	0.5	1.1
	954	0.7	
2-methylpentane	1039	0.4	1.2
	1147	0.6	
2-methylhexane	1144	0.6	2.5
	1304	0.74	
Ethyl alcohol	1040	0.52	1.7

TABLE 5. Comparison of Times τ_0 and τ_r

Substance	Mean orient., $\tau_0 \cdot 10^{12}$, sec	Mean relaxation, $\tau_0 \cdot 10^{12}$, sec	
		data from [7]	data from [10]
Benzene	2.0	3.3	2.0
Toluene	3.0	5.3	4.2
Paraxylene.	4.0	—	9.6
Metaxylene	4.3	—	6.6

Table 5 facilitates a comparison of the values we measured for the mean reorientation time τ_0 using the data acquired from measurements of the relaxation time τ_r of anisotropy fluctuations reported by other authors [7, 10] for the adjacent portion of the wind of the Rayleigh line. The reader will readily see from Table 5 that the times τ_0 and τ_r agree as to order of magnitude. The absolute values of the times τ_0 and τ_r in the case of benzene are quite close, but differ by factor of from 1.5 to 2.5 in the case of toluene, metaxylene, and paraxylene.

The temperature dependence of the mean reorientation time of the molecule was predicted theoretically by Ya. I. Frenkel', and is described by the formula (I.1)

$$\tau_{or} = \tau_0 = \tau' \exp (U_{or}/kT).$$

This formula contains one other important parameter specifying the Brownian rotatory motion of the molecules: the reorientation potential barrier U_{or} of a molecule in the liquid. The potential barriers U_{or} may be determined with the aid of our experimental data on the basis of the theoretical concepts put forth by Ya. I. Frenkel'.

Making use of the formulas

$$\tau_0 = \tau' \exp (U_{or}/kT) \quad \text{and} \quad \Delta\nu = 1/\pi c \tau_0.$$

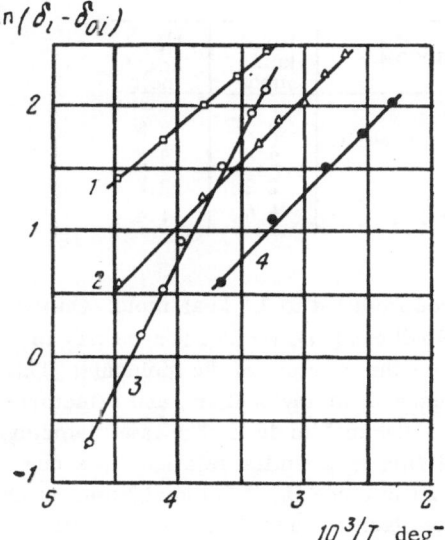

Fig. 13. Dependence of $\ln(\delta_i^- - \delta_{0i})$ on $1/T$. 1) Isopentane ($\nu = 909$ cm^{-1}) in the temperature range from $-50°$ to $30°$C; 2) toluene ($\nu = 623$ cm^{-1}) from -50 to $100°$C; 3) ethyl alcohol ($\nu = 1040$ cm^{-1}) from $-60°$ to $30°$C; 4) metaxylene ($\nu = 517$ cm^{-1}) from $0°$ to $160°$C.

and remembering that

$$\delta_i\,(T) = \delta_{0i} + \Delta\nu\,(T),$$

we obtain the following formula for the width of the depolarized Raman line:

$$\delta_i\,(T) = \delta_{0i} + A\,\exp\,(-\,U_{\mathrm{or}}/kT_i), \qquad (\mathrm{II.\,7})$$

where $A \simeq 1/\pi c\tau'$. Setting up a system of equations for the three distinct temperatures:

$$\begin{aligned}
\delta_i\,(T_1) &= \delta_{0i} + A\,\exp\,(-\,U_{\mathrm{or}}/kT_1), \\
\delta_i\,(T_2) &= \delta_{0i} + A\,\exp\,(-\,U_{\mathrm{or}}/kT_2), \qquad (\mathrm{II.\,8}) \\
\delta_i\,(T_3) &= \delta_{0i} + A\,\exp\,(-\,U_{\mathrm{or}}/kT_3),
\end{aligned}$$

we are now in a position to compute U_{or}. The solution of the system of equations (II.8) reduces to the solution of the transcendent equation

$$\frac{\delta_i\,(T_1) - \delta_i\,(T_2)}{\delta_i\,(T_1) - \delta_i\,(T_3)} = \frac{\exp\,(-\,U_{\mathrm{or}}/kT_1) - \exp\,(-\,U_{\mathrm{or}}/kT_2)}{\exp\,(-\,U_{\mathrm{or}}/kT_1) - \exp\,(-\,U_{\mathrm{or}}/kT_3)}, \quad (\mathrm{II.\,9})$$

which is reducible to a form well suited to graphical solutions. In that case, there will be no need to determine the "residual width."

In our work, the problem of determining the potential barriers was measurably simplified in that we studied the temperature dependence of the line width over a broad range of values. From the dependence of the Raman line width on a quantity reciprocal to the viscosity, we found the "residual width" δ_{0i} by extrapolation, and then plotted a graph $\ln[\delta_i(T) - \delta_{0i}] = f(1/T)$. After taking the logarithm of Eq. (II.7), we realize that were the potential barrier U_{or} to be a function of the temperature, straight lines would be plotted on the graph and U_{or} could then be determined from the slopes of those lines.

As an illustration, Fig. 13 shows graphs of $\ln[\delta_i(T) - \delta_{0i}] = f(1/T)$ for isopentane over the range of values from $-50°$ to $30°$C, for toluene over the range from $-50°$ to $100°$C, for ethyl alcohol for the range from $-60°$ to $+30°$C, and for metaxylene over the range from $0°$ to $160°$C. As we readily see from the graphs, the experimental data points fit well on the straight line corresponding to each substance. This is also indicative of the fact that the potential barrier U_{or} is independent of the temperature over the range studied. The barrier was calculated for each liquid on the basis of several depolarized Raman lines belonging to different vibrations of the molecule. Within the limits of experimental error, the value of U_{or} was found to be independent of the line selected, and is consequently independent of the mode of vibration of the molecule.

Below we list values of the potential barriers calculated for ten different liquids. The mean error did not surpass ± 0.5 kcal/mole.

Substance	U_{or}, kcal/mole	Substance	U_{or}, kcal/mole
Isopentane	1.6	Benzene	2.4
2-methylpentane	2.0	Toluene	2.3
2-methylhexane.....	2.5	Metaxylene.........	2.2
Cyclopentane	1.4	Paraxylene	2.2
Cyclohexane	2.8	Ethyl alcohol........	3.6

TABLE 6. Values of Potential Barriers

Substance	U_{or}, kcal/mole	U_b, kcal/mole	Substance	U_{or}, kcal/mole	U_b, kcal/mole
Isopentane	1.8	1.6	Toluene	2.2	2.3
2-methylhexane . .	2.2	2.5	Metaxylene	2.4	2.2
Ethyl alcohol	3.6	3.6	Paraxylene.	2.2	2.2
Benzene.	2.4	2.4	Cyclopentane	1.6	1.4

The value of the potential barriers of the substances studied ranged from 1.4 to 3.6 kcal/mole. On the basis of our results and data reported from measurements by A. L. Sokolovskaya [27], we feel it necessary to remark that, as a rule, the potential barrier U_{or} will be lower, the simpler the structure of the molecule. This is quite clear against the example of measurements referable to isopentane, 2-methylpentane, and 2-methylhexane. Isopentane exhibits the simplest molecular structure, and U_{or} = 1.6 kcal/mole in that case; 2-methylhexane exhibits a more complex molecular structure and U_{or} = 2.5 kcal/mole; a similar relationship is observed to occur in the case of cyclopentane and cyclohexane. In the first instance U_{or} = 1.4 kcal/mole; in the second U_{or} = 2.8 kcal/mole. The practically identical potential barrier observed in the case of molecular species in the aromatic series is probably due to the structure of the benzene ring, which lies at the basis of all four molecular species.

The highest value of U_{or} of those species studied is that of ethyl alcohol, apparently because of the strong intermolecular interaction brought about by the presence of the hydrogen bond.

The constant value of the potential barrier U_{or} detected at various temperatures of the liquids studied (cf. Fig. 13) suggests that the immediate surroundings of a molecule, which are a deciding factor in the value of U_{or}, are not altered as the temperature of the liquid changes. After each time the molecule flips over, and independently of the temperature, the number of nearest neighbors, interaction with which will affect the probability that the molecule will flip, remains the same.

It should be noted that Ya. L. Frenkel' [4], H. Eyring [28], and others derived a relationship linking the viscosity to the potential barrier:

$$\eta = \text{const} \cdot e^{\frac{U_b}{kT}} , \qquad (II.10)$$

where k is the Boltzmann constant, T is the absolute temperature, and U_b is the potential barrier which a molecule of the liquid has to climb over in the transition from one equilibrium position to another. This formula is being widely used in molecular physics to determine U_b (e.g., cf. reference [29]).

Since the potential barrier U_{or}, measured in terms of the line width, characterizes the transition of a molecule from one equilibrium position to another by reorientation, a comparison of U_{or} and U_b affords some information on the nature of the motion of molecules occurring in the liquid and responsible for the viscous flow.

Table 6 lists the values of U_b (data on viscosity taken from the reference books [25, 26]) which we computed and the values of U_{or} calculated from the temperature dependence of the width of depolarized Raman spectral lines. We see readily from the table that the potential barriers agree within the limits of error of the measurements. This agreement could not be coincidental, for it is observed in many substances belonging to the most varied classes of chemical compounds.

The results arrived at are of enormous significance for any understanding of the nature of viscous flow, and apparently indicate that jumps by molecules from one equilibrium position to another and responsible for viscosity will occur when the molecules reverse their orientations. M. F. Vuks [30] arrives at a similar conclusion. He computed molecular reorientation potential barriers U_{or} for many liquids from Rayleigh scattering data. Comparison of the values so obtained and the U_b barriers obtained from viscosity data demonstrated that U_{or} = U_b, as a rule. On this basis, that author suggests that a molecule jumps from one equilibrium position to another simultaneously with a reversal of its orientation.

It is evident from the material cited that the temperature dependence of the width of depolarized Raman lines may be useful for determining the mean reorientation time τ_0 of a molecule in the liquid. It has been found that τ_0 increases as the structure of a molecule of a species belonging to a common homologous series becomes more complex. The computed τ_0 times turn out to be equal in order of magnitude to the relaxation times for anisotropy fluctuations as determined from the near side of the wing of the Rayleigh line. Molecular reorientation potential barriers U_{or} in a liquid were measured, on the basis of Raman line width, for a large number of molecular species. It was found that, as a rule, a simpler molecular structure means a lower value of U_b. It was also demonstrated experimentally that the barriers U_{or} agree with the potential barriers U_b arrived at on the basis of viscosity data. This result is of paramount importance in understanding the nature of viscous liquid flow, and is apparently evidence in favor of the view that jump of a molecule from one equilibrium position to another is accompanied by a reversal of the molecule's orientation.

CHAPTER III

INVESTIGATION OF THE ROTATORY MOTION
OF MOLECULES IN CRYSTALS BY THE RAMAN
SCATTERING METHOD

§1. Nature of the Rotatory Motion of Molecules in Crystals

The nature of molecular reorientations in crystalline substances is determined by the symmetry properties of the molecules and of the crystal lattice. The lattice symmetry must not be impaired as the molecule reverses its orientation inside the crystal. Hence, the molecules can undergo reorientation inside the crystal lattice only with respect to their symmetry axes, or with respect to the symmetry axes of the unit cell. Depending on the lattice structure and on the symmetry properties of the molecule, the probability of a molecular reorientation with respect to the distinct axes of the lattice will be different in each case.

Reorientation of molecules inside the crystal lattice, with respect to the axes of symmetry, was observed without any accompanying change in the centers of gravity of the molecules in researches conducted by the nuclear magnetic resonance method [31]. It was shown in that monograph that the reorientation probability of a molecule in a lattice may be expressed as

$$w = \text{const} \cdot e^{-\frac{U_{cr}}{RT}},$$ (III.1)

where U_{cr} is the reorientation potential barrier of the molecule in the lattice. These molecular reorientations in crystalline substances must exert an appreciable influence on the width of the depolarized Raman lines. The width of the Raman lines, influenced, as pointed out earlier, by the rotatory movement of the molecules, may be described in this case by the formula

$$\Delta \nu_{cr} = \frac{1}{\pi c \tau_{cr}},$$ (III.2)

where τ_{cr} is the mean reorientation time of the molecule in the lattice. Allowing for the fact that the reorientation probability of the molecule in the lattice, w, is related to the time τ_{cr} by the formula

$$w = 1/\tau_{cr},$$ (III.3)

and bearing Eqs. (III.1) and (III.2) in mind, we obtain

$$\Delta \nu_{cr} = \frac{1}{\pi c \tau_{cr}} = A \exp\left(-\frac{U_{cr}}{RT}\right).$$ (III.4)

Consequently, the temperature variation of the width of the depolarized Raman lines may yield information on the mean reorientation time and on the reorientation potential barrier of the molecule in the crystal lattice, and the Raman scattering method may be used in the study of the rotatory motion of molecules within crystalline substances.

Since the probability that the molecule's reorientation with respect to the several axes of symmetry may be different, several cases are possible where molecular reorientation will be most probable only with respect to one particular axis of symmetry. If the induced moment of some mode of molecular vibration whose tensor of the derivative polarizability with respect to the normal coordinate has an anisotropic part lines up in that case with the axis of reorientation, the projections of the amplitude of the scattered light wave on the coordinate axes of observations will not vary in time. In accord with theory, the width of the line corresponding to that mode of vibration will be independent of the rotatory motion of the molecules, and will be likewise independent, as a consequence, of the temperature of the crystal. In this context it does not matter whether we are dealing with a single crystal or a polycrystal.

If a reorientation of the molecule in the crystal is equally likely about two or more symmetry axes, then the width of all the depolarized Raman lines must be a function of the crystal temperature, inasmuch as even when the induced moment of some one mode of vibration does line up with one of the symmetry axes, its projections onto the coordinate axes of the observer's system will vary as the molecule reorients with respect to the other axes of symmetry.

The arguments adduced above allow us to formulate the following rule of thumb: if we find in the spectrum of the crystalline host substance, along with depolarized Raman lines whose width is a function of the temperature of the substance, so much as one single depolarized Raman line whose width remains constant and indifferent to temperature variations, then the molecule will be most likely to reorient, inside the crystal lattice, about a single axis of symmetry and only about that one. The induced moment of the molecular vibration to which the Raman line of temperature-independent width corresponds will then line up with the direction of that axis of reorientation.

As mentioned earlier, a molecular reorientation inside the crystal lattice may even occur relative to the symmetry axes of the unit cell, which can in turn exert its own influence on the Raman line width. In case of molecular crystals, there are two or more molecules in a unit cell and the symmetry axes of the unit cell are not in coincidence with the symmetry axes of the molecule [32]. Because of this, reorientation of the molecule relative to the axes of the unit cell will occur consonant with a change in the position of the center of gravity. This displacement of the molecule within the crystal lattice is due, according to the concepts advanced by Ya. I. Frenkel', to the presence of "vacancies" or "holes" in the host lattice, and the probability of the molecule reorienting relative to the axes of the unit cell is equal to the probability of a "hole" forming. The probability of a "hole" forming is in turn equal to the probability that a molecule of the crystal will evaporate, and may be stated as

$$w' = \text{const} \cdot e^{-\frac{U_{evap}}{kT}}, \tag{III.5}$$

where U_{evap} is the latent heat of evaporation. An estimate which we made indicated that the probability of the molecule reorienting with respect to the symmetry axes of the unit cell is many times less than the probability of the molecule reorienting with respect to its own axes of symmetry. This means that the reorientation of molecules relative to the axes of the unit cell will exert practically no effect whatever on the width of the depolarized Raman lines.

The discussion developed in this section demonstrates that the Raman scattering method is valuable research tool in studying the nature of random flips of the molecules within a crystal lattice, via determinations of the mean reorientation time of the molecule, the corresponding potential barrier, and the axes of symmetry of the molecule about which molecular reorientation is most probable.

§ 2. Experimental Research on Molecular Reorientations in the Crystal Lattice

In selecting research objects for the study of the rotatory motion of molecules within the crystal lattice, we took as our point of departure the following considerations. First, the molecules of the crystals under investigation must exhibit symmetry axes of not lower than second order. Second, it is advisable to investigate the reorientations of molecules having a different number of axes of symmetry of different orders. And thirdly,

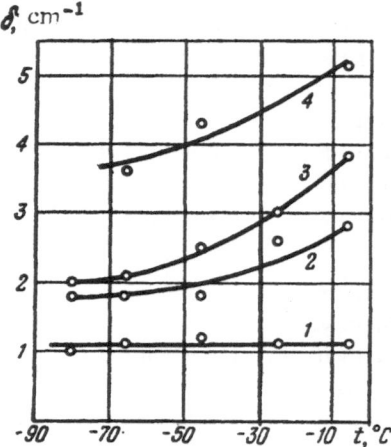

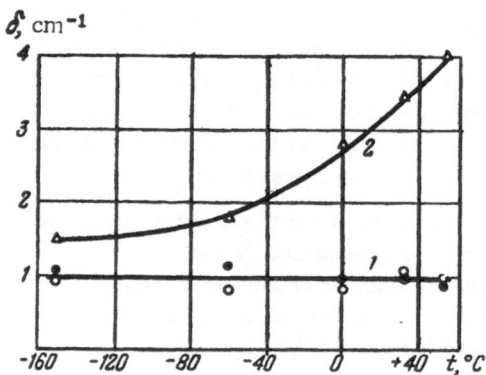

Fig. 14. Temperature dependence of width of Raman spectral lines of crystalline p-xylene. 1) ν = 645 cm^{-1}, ρ = 0.89; 2) ν = 829 cm^{-1}, ρ = 0.14; 3) ν = 1618 cm^{-1}, ρ = 0.72; 4) ν = 3012 cm^{-1}.

Fig. 15. Temperature dependence of width of Raman spectral lines of crystalline p-dichlorobenzene. 1) ν = 307 cm^{-1}, ρ = 0.86 (black circles), ν = 627 cm^{-1}, ρ = 0.86 (hollow circles); 2) ν = 1578 cm^{-1}, ρ = 0.6.

the melting point of the crystals under investigation must be comparatively high (270° to 300°K) in order for the width of the lines, dictated by the molecular reorientation, to have the dimension inferred from Eq. (III.4).

Because of this, we made the choice of the following substances in our investigation: paraxylene (melting point 13.2°C), the molecule of which has only one twofold axis of symmetry; paradichlorobenzene (melting point 52.7°C), the molecule of which has three twofold axes of symmetry, and cyclohexane (melting point 6.5°C), the melting point of which has three twofold axes of symmetry and one threefold axis of symmetry. These substances were studied in the polycrystalline state. The widths of lines of different depolarization ratio ρ were measured over a broad range of temperatures in the spectra of the crystal investigated. The results of the determinations of the temperature dependence of Raman line width in these instances appear plotted in Figs. 14 to 16.

The measurements performed revealed that the line widths in the case of low ρ values depend very slightly on the crystal temperature. The widths of some of the depolarized Raman lines ($\rho \approx ^6/_7$) vary appreciably in response to a temperature change, narrowing down gradually as the crystal temperature is lowered and tending to some constant in the case of each specific Raman line. This behavior of Raman line width as a function of the temperature bears a resemblance to the temperature behavior of the line width in the spectra of liquids.

However, in the spectra of paraxylene and paradichlorobenzene crystals the depolarized Raman lines ν = 645 cm^{-1} (paraxylene), ν = 307 cm^{-1}, and 627 cm^{-1} (paradichlorobenzene), whose widths remain virtually constant over the entire range of variation in crystal temperatures, were discovered. There are solid grounds validating the notion that the temperature independence of the width of these lines is due to the above properties characterizing the reorientation of them the molecules in the crystal lattice.

Let us consider some of the findings of the measurements in regard to each substance in particular.

Paraxylene (Melting Point 13.2°C). The spectrum of paraxylene was studied over the temperature range extending from −5° to −80°C. There is one depolarized Raman line ν = 645 cm^{-1} in the spectrum of this crystal (Fig. 14), with a width which is independent of the temperature of the substance. The widths of the two other Raman lines, ν = 1618 cm^{-1} and 3012 cm^{-1}, vary with the temperature. The experimentally detected pattern of variation in the width of the lines referred to with the temperature allow us to infer that the induced moment of the ν = 645 cm^{-1} vibration is aligned with the reorientation axis of the

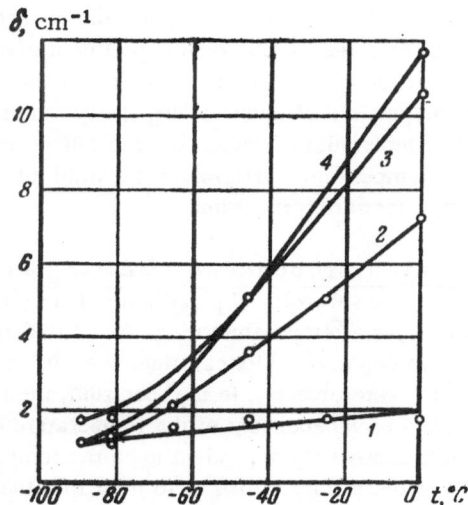

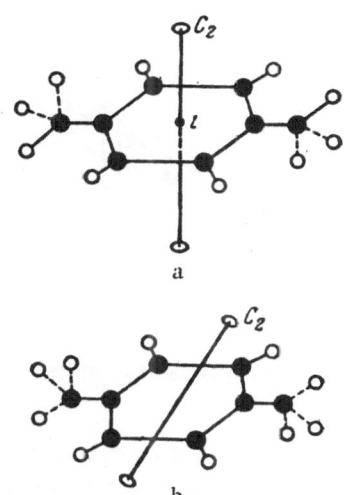

Fig. 16. Temperature dependence of width of Raman spectral lines of crystalline cyclohexane. 1) $\nu = 802$ cm^{-1}, $\rho = 0.13$; 2) $\nu = 1029$ cm^{-1}, $\rho = 0.79$; 3) $\nu = 1445$ cm^{-1}, $\rho = 0.81$; 4) $\nu = 1267$ cm^{-1}, $\rho = 0.77$.

Fig. 17. Structure of the p-xylene molecule. a) C_{2h}; b) C_{2v}. Black dots indicate carbon atoms, and hollow circles denote hydrogen atoms.

molecule. Taking into account the symmetry properties of the paraxylene molecule, we are in a position to attempt to find its reorientation axis in the crystal lattice.

The membership of the paraxylene in a specific point group of symmetry C_{2h} or C_{2v} is obscured by the fact that data are lacking on the configuration of the methyl groups. In Fig. 17 we have diagrams of the structure of the paraxylene molecule corresponding to the symmetry groups above. In the case of the C_{2h} group, the H_1 and H_2 atoms are to be found in the same plane as the benzene ring (Fig. 17a). The remaining hydrogen atoms are noncoplanar. In this configuration, the molecule possesses a center of symmetry i and a single twofold axis of symmetry perpendicular to the plane of the benzene ring. In the case of the C_{2v} group, all the hydrogen atoms belonging to the methyl groups lie out of the plane in which the benzene ring is situated (Fig. 17b). In this case, the molecule does not have a center of symmetry, and has only one twofold symmetry axis.

The results obtained from measurements of the temperature variation of the width of lines of the Raman spectrum of paraxylene permit us to find out which of the two symmetry groups mentioned is the most likely candidate for the crystalline paraxylene molecule. The Raman line $\nu = 645$ cm^{-1}, whose width remains constant as the crystal temperature varies, corresponds to an in-plane vibration of the molecule [33]. According to the selection rules, the induced moment of these vibrations will lie in the plane of the benzene ring. In the case of the C_{2h} symmetry group, reorientation of the molecule is most probable only about the C_2 axis perpendicular to the benzene ring. The induced moment will then change its direction as the molecule reorients, and consequently the amplitude of the scattered light wave will be modulated by the rotatory motion. This implies that the width of the Raman line $\nu = 645$ cm^{-1} must be a function of the crystal temperature, but this is not observed experimentally. The experimental data thus stand in contradiction to the model of the paraxylene molecule corresponding to the symmetry group C_{2h}.

In the case of the symmetry group C_{2v}, the reorientation axis and the induced moment of the $\nu = 645$ cm^{-1} vibration lie in the same plane. For that reason, we may not exclude a possible coincidence in the direction of the induced moment of that vibration of the molecule and the reorientation axis. This means that we may surmise that the molecule of crystalline paraxylene belongs to the C_{2v} symmetry group. The lines $\nu = 1618$ cm^{-1} and $\nu = 3012$ cm^{-1} do not coincide with the induced moment of the $\nu = 645$ cm^{-1} vibration, and accordingly do not coincide with the reorientation axis of the molecule either. The width of these

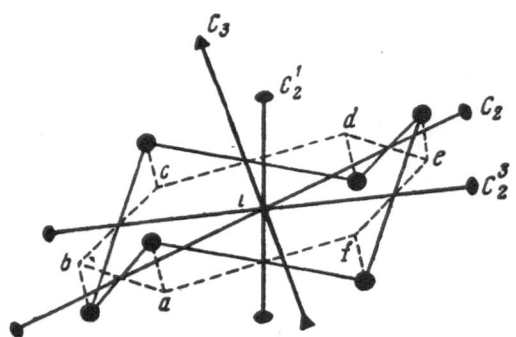

Fig. 18. Structure of the cyclohexane molecule.

Raman lines must accordingly vary with the changes in temperature, as is indeed observed experimentally. We note here that the case in point furnishes a clearcut example of the possible of investigating the symmetry properties of molecules of crystalline substances on the basis of the temperature variation of the width of lines in the Raman scattering spectrum.

Paradichlorobenzene (Melting Point 52.7°C). The spectrum of paradichlorobenzene was investigated over a temperature range from +52° to −150°C. Two depolarized Raman lines $\nu = 307$ cm^{-1} and 627 cm^{-1} were observed in the spectrum, their widths being found to be independent of the temperature of the substance, and the line $\nu = 1578$ cm^{-1} was also found, with a width sensitively dependent upon the temperature (Fig. 15). The pattern of variation discovered in the width of these lines allows us to infer that the reorientation of the paradichlorobenzene molecule in the crystal lattice is most likely to occur about one of its the symmetry axes of the lattice. Once the symmetry axes are known and assignments are made to the frequencies of the paradichlorobenzene Raman spectrum, we are in a position to draw certain conclusions regarding which of the symmetry axes of the molecule is the axis of reorientation in the crystal lattice.

The paradichlorobenzene molecule belongs to the point group of symmetry $V_h = D_{2h}$ and possesses three twofold axes of symmetry. Two of these axes lie coplanar with the benzene ring, one passing through the chlorine axes and the other perpendicular to the first. The third axis passes through the center of symmetry i of the molecule and runs perpendicular to the plane of the benzene ring. Utilizing data in the literature [34], where the frequency assignments are made in the paradideuterobenzene spectrum, we can now show that both depolarized Raman lines with widths independent of the crystal temperature belong to the in-plane vibrations of a molecule in the B_{1g} symmetry class.

Since these Raman lines belong to in-plane vibrations, the induced moments of these vibrations lie, according to the selection rules applying, in the plane of the benzene ring and their directions cannot possibly coincide with the symmetry axis perpendicular to the benzene ring. Reorientation of the molecule about that axis is highly improbable for that reason. Unfortunately, it is not now possible to determine exactly precisely about which of the other two axes lying in the plane of the benzene ring reorientation will be most probable. It might be assumed that the most probable axis of reorientation is the symmetry axis passing through the chlorine atoms, since the moment of inertia of the molecule about this axis is almost one order of magnitude less than the moment of inertia of the molecule about the second axis lying in the plane of the benzene ring.

Cyclohexane (Melting Point 6.5°C). The spectrum of crystalline cyclohexane was studied over a range of temperatures from 0° to −90°C. M. M. Sushchinskii [35] has calculated the frequencies and made an interpretation of the Raman spectrum. Selection rules resolve the symmetric vibrations of the A_{1g} symmetry class and the doubly degenerate oscillations of the E_g symmetry class in the Raman scattering spectrum. For the study of the reorientations of the molecule in the crystal lattice, we have to choose the lines corresponding to vibrations of the E_g class, since the tensor of the polarizability derivative with respect to the normal coordinate is anisotropic in the case of these vibrations. Calculations indicate that eight such modes of vibration should be manifested in the spectrum. Seven Raman frequencies have been observed experimentally to date.

We have determined the temperature dependence of the width of the Raman lines 802 cm^{-1} (A_{1g} class), 1029, 1267, 1445, and 2852 cm^{-1} (E_g glass). The lines enumerated are the most intense lines in the spectrum and are suitable for width measurements. As the crystal temperature varies, the width of the Raman line 802 cm^{-1} varies very slightly, while the width of lines belonging to the class E_g of molecular vibrations vary quite pronouncedly (Fig. 16), indicating the extent of the effect of the rotatory motion of the molecules in the crystal lattice.

The cyclohexane molecule belongs to the point group of symmetry D_{3d} and has a single threefold axis of symmetry and three twofold axes of symmetry. Figure 18 shows the relative positions of the molecule's symmetry axes. The twofold symmetry axes lie in the abcdef plane passing lying halfway between the upper and lower atoms of the chair-ring skeleton of the molecule, and make 120° angles with each other. The threefold axis is at right angles to that plane and passes through the center of symmetry of the molecule.

Andrew and Eades [36] studied crystalline cyclohexane by the nuclear magnetic resonance method. These authors demonstrated that reorientation of the cyclohexane molecule in the crystal lattice, over the temperature range from +6.5° to −87°C, is equally probable about any of the three twofold axes of symmetry. It is precisely this reorientation of the cyclohexane molecules which exerts a substantial effect on the width of Raman lines corresponding to vibrations of the molecule in the E_g class.

Temperature investigations of the dependence of the width of depolarized lines in the Raman spectra of crystalline substances enable us to determine the potential barriers U_{cr} hindering reorientation of the molecule in the lattice. The determination of U_{cr} and the comparison of this barrier height with the potential barrier to reorientation of a molecule in a liquid affords an opportunity to study the effect of the environment of a molecule on the reorientation probability of the molecule in liquids and in crystals. In a crystalline substance the surroundings of the molecule are virtually constant, and are characterized by a certain order and symmetry related to the properties of the crystal lattice.

The ordering of the molecules varies appreciably in the transition from the crystalline to the liquid state. Long-range ordering is absent in the structure of a liquid. The problem of close-range ordering in a liquid has been subjected to detailed analysis in Fisher's monograph [1]. According to the concepts advanced by that author, if a small number of neighbors (say 10 to 20) are singled out in the environs of a molecule, then this group of particles may be found to possess order over a span of time of the order of the mean lifetime of the molecule in an equilibrium position. As the time passes these ordered groups will be broken up and new ones will be formed. The lifetime of these groups will shorten rapidly as the temperature of the liquid increases. In the theory, these groups have been dubbed "close-order groups."

If the "close-order" structure exerts an effect on the potential barrier hindering reorientation of the molecule, then the heights of the barrier should be different in a liquid and in a crystal. We list below some experimentally derived values of the potential barriers (in kcal/mole) to reorientation of paraxylene and cyclohexane molecules in the liquid state U_{or} and in the crystal state U_{cr}, measured on the basis of the temperature dependence of the Raman line widths. It is clear from the data below that the potential barriers hindering reorientation of the molecules of these molecular species are the same in both phases of matter for the two species, within the limits of experimental error.

Molecular species	Crystalline phase	Liquid phase
Paraxylene	3.0 ± 0.5	2.2 ± 0.5
Cyclohexane	2.8 ± 0.5	2.8 ± 0.5

The constant value of the potential barrier heights detected in the liquid and crystalline phases are apparently an indication that the environment of the molecule in each molecular species remains more or less the same in both phases. This in turn permits us the assumption that the closest neighbors lying on the first coordination sphere exert their effect on the reorientation of the molecule. According to Fisher [1], the first coordination number in the transition from a liquid to a crystal will either increase slightly or else will remain the same. As a result, the potential barrier to reorientation in the transition from the liquid to the crystal state will likewise either not change at all or else will raise its height only slightly. This increase in the height of the potential barrier is observed, for instance, in the case of benzene. The potential barrier for liquid benzene which we measured showed a height of 2.4 kcal/mole. The potential barrier to reorientation of the benzene molecule in a crystal lattice, as measured by the nuclear magnetic resonance method [31], was found to be 3.6 kcal/mole.

The material cited in this chapter demonstrates that an investigation of the temperature variation of the width of Raman spectral lines of crystalline substances provides us with more precise information on the symmetry of the molecules and aids in our study of the special features of the rotatory motion of molecules in a crystal lattice, so that we can determine the height of the potential barrier and the mean reorientation time of the molecule in the lattice, as well as the most probable axes of reorientation of the molecules.

CHAPTER IV

INVESTIGATION OF THE REORIENTATIONS
OF MOLECULES IN LIQUID
AND CRYSTALLINE SUBSTANCES
BY THE INFRARED ABSORPTION METHOD

§1. Equipment and Procedure for Measuring the Width of Infrared Absorption Bands

Infrared absorption spectra were recorded on a double-beam automatic IKS-14 spectrometer using a NaCl prism. The infrared radiation sensor was a metal bolometer fabricated at our own laboratory [37]. The instrument was calibrated in accordance with the indene spectrum [38]. The bands under investigations were recorded at constant spectrometer slit width. The slit openings were selected in each case such that the monochromator slits would introduce minimal distortion into the band profile. As the band contours were being recorded the scanning speed was set at 1.8 cm^{-1}/min. The rate of feed of the strip-chart paper was set at 6 mm/min.

Constant-thickness IR cells made of crystalline NaCl plates were utilized in recording the infrared absorption spectra. To eliminate reflections from the windows of the cell, measurements were carried out with two IR cells of different thickness placed in the two light beams of the instrument. Cells containing the substance under investigation were placed in a special thermostat in which the temperature could be varied subject to control. A detailed description of the thermostating device will be found in reference [39].

Some knowledge of the apparatus function of the instrument is required in measurements of the true width and shape of the IR absorption bands. Determining the apparatus function of infrared spectrometers is no easy task. It is important to note that even at the present time apparatus functions have not yet been measured for the most commonly used standard instruments.

We made use of data reported by G. G. Petrash [40] in finding the apparatus function of the IKS-14 IR spectrometer which we employed in our studies. That author measured the shape and width of several infrared absorption bands, using a double-beam IR spectrometer with diffraction grid, described in [41], in his work. The instrument enabled the author to work with slits 0.4-0.7 cm^{-1} at a slow scanning rate, making it possible to record all of the bands under study practically free of systematic distortions. Of the seven IR absorption bands studied by G. G. Petrash, two exhibited a shape close to the dispersion curve, both in optical density and in absorption: the 903 cm^{-1} band in the cyclohexane IR spectrum and the 1147 cm^{-1} band in the pyridine spectrum. The remaining bands exhibited pronounced asymmetry or else showed weak subsidiary maxima, so that their shape was not amenable to description by any simple analytic formula. The measured true width of the 903 cm^{-1} IR spectral band of cyclohexane was 4.5 cm^{-1} (in optical density), while that of the pyridine 1147 cm^{-1} band was 6.9 cm^{-1}. These two bands were the ones used in the determination of the apparatus function of the IKS-14 spectrometer.

The apparatus function was measured in the following manner. The 903 cm^{-1} band of cyclohexane and the 1147 cm^{-1} band of pyridine were recorded at the constant monochromator slit widths S = 2.0 cm^{-1} for the first band and S = 3.0 cm^{-1} for the second band. Total transmission bands and 100% absorption bands were recorded, during this work, on each spectrogram. The profile shape and the width of the bands observed in absorption were also measured. It was found that the shape of the observed IR bands is described by a dispersion type curve, in accord with the findings reported by Petrash [40]. The assumption in the calculation of the width of the apparatus function was that the monochromator apparatus function has a Gaussian shape. The width of the apparatus function was determined from the known width of the observed profile and the true width by using the graph of the convolution of the Gaussian and dispersion curves (cf. Fig. 2). As a result of the computations, it was discovered that the width of the apparatus function was equal to the spectral width of the spectrometer widths within the limits of error of the measurements. In our case, accordingly, the apparatus function of the instrument was found to be, for all practical purposes, the same as the slit apparatus function, and elimination of that function from the observed results involves no particular difficulties.

As a result, the method used to measure the true width of the investigated IR bands reduced to the following. The band under study and the total transmission and 100% absorption bands were recorded on each spectrogram. The profile of the absorption band was reduced (in absorption) to an infinitely narrow slit by Rayleigh's graphical method, which was described in Chapter II. The band profile so corrected was converted to optical density, and the true width of the IR band was measured later. All of the results of measurement of the δ_d bands cited in this article are given for the profiles of bands in optical density. The width of each band was measured from three to four spectrograms; the mean error was ± 0.5 cm^{-1}.

In selecting the objects of investigation with the object of shedding light on the nature of the rotatory motion of molecules, the following factors are to be taken into account. In the IR absorption bands, there must be present some solitary bands corresponding to the fundamentals, and not coinciding with the overtones. Moreover, the intermolecular coupling, according to [32], may result in broadening of the band and even in splitting of the band, if this vibration is both Raman-active and active in infrared absorption. For that reason we should select bands which do not appear in the Raman spectrum. Note here that these conditions add extra difficulties to the choice of objects, since such bands are actually rare in the spectra.

§2. Investigation of the Dependence of Bandwidth on Aggregate State, Temperature, and Viscosity

In order to make a study of the influence of the rotatory motion of molecules in liquids and solids, we selected the following compounds: cyclohexane, toluene, and acetonitrile. Let us consider some of the results reported in the literature for each of these species in particular.

Cyclohexane. Petrash [35] presented calculations and an interpretation of the Raman scattering spectrum and infrared absorption spectrum of cyclohexane. On the basis of the data cited in that paper, we chose the single band 903 cm^{-1} in the IR spectrum, corresponding to the fundamental vibration of a molecule belonging to the A_{2u} symmetry class, in our studies. The shape of the profile of this band was studied earlier by G. G. Petrash [40], both in absorption and in optical density. The temperature dependence of the width of the depolarized Raman lines of liquid cyclohexane had been studied elsewhere [21], and that of the crystalline state was discussed in the account of our present investigations (cf. Chapter III).

According to the selection rules, the dipole moment of the 903 cm^{-1} vibration of the cyclohexane molecule is not aligned with the direction of any of three principal axes of the molecule, while the bandwidth of that vibration must be a function of the temperature of the substance. The results of a measurement of the width δ_d of that band in the temperature range from $+60°$ to $-60°C$ are plotted in Fig. 19. The study of the profiles of this band at various temperatures of both liquid and crystalline cyclohexane showed that the observed profiles have a shape close to that of a dispersion profile.

As we see from Fig. 19, the width of the band under investigation narrows down as the temperature of the substance is lowered. In the transition to the crystal (melting point $T_g = 6.5°C$) no abrupt changes are observed in the width of the band. This nature of the temperature dependence of the width of the cyclohexane IR spectral band is analogous to the pattern of temperature behavior of its Raman line widths. A similar change in the width of bands in the infrared absorption spectra was reported by Lisitsa et al. [42], who studied the spectra of liquid and crystalline carbon tetrabromide.

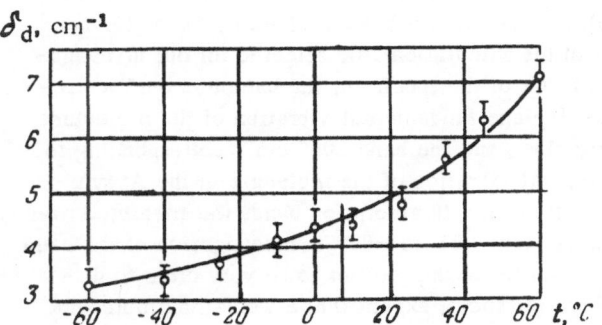

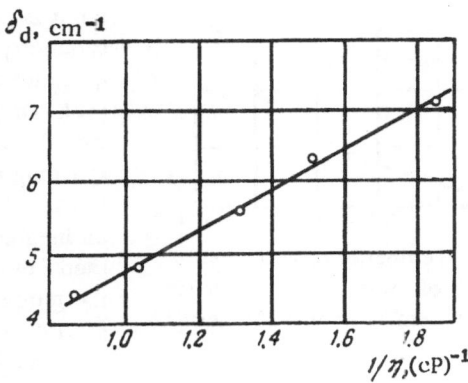

Fig. 19. Temperature dependence of the width of the 903 cm^{-1} band in the IR spectrum of cyclohexane.

Fig. 20. Width of the 903 cm^{-1} spectral band of cyclohexane as a function of $1/\eta$.

Our measurements and the measurements reported in reference [42] showed the width of the absorption bands to narrow with a decrease in temperature, tending to some constant value. On that basis, we infer that the width of the absorption band constitutes a sum of the width $\Delta \nu(T)$ due to the rotatory motion of the molecules and the width δ_0 which is independent of the rotations of the molecules. This quantity is what was given the name "residual width" in the Raman scattering case. The width of the absorption band, just as the width of the Raman lines, may be specified, therefore, by the equation

$$\delta(T) \approx \delta_0 + \Delta \nu(T). \tag{IV.1}$$

Pursuing our arguments along the same lines as in the Raman scattering case (cf. Chapter II, this article), and taking due cognizance of the fact that the width of the band due to the rotatory movement of the molecule is

$$\Delta \nu(T) = 1/\pi c \tau_0 T,$$

we obtain the following expression for the total width of the IR absorption band:

$$\delta(T) \simeq \delta_0 + A \exp(-U_{or}/kT). \tag{IV.2}$$

Our experimentally derived data on the temperature behavior of the 903 cm^{-1} absorption band enable us to determine the height of the reorientation potential barrier U_{or} of the molecule. The U_{or} values found for liquid and crystalline cyclohexane fit well with the measurements carried out on the Raman scattering spectrum. Below, the reader will find the barrier heights U_{or} (kcal/mole) obtained by the two methods, for comparison.

Molecular species	Raman scattering	Infrared absorption
Cyclohexane (liquid)	2.8 ± 0.5	2.9 ± 0.5
Cyclohexane (crystal)	2.8 ± 0.5	2.9 ± 0.5

It was shown back in Chapter II that the width of depolarized Raman lines of pure liquids constitutes a linear function of the reciprocal viscosity. If the width of the absorption band is due more than anything else to the rotatory Brownian movement of the molecules, then it would be natural to expect that its width would also be a linear function of $1/\eta$ in the case of infrared absorption bands as well. Figure 20 shows a graph plotting the width of the 903 cm^{-1} IR absorption band as a function of $1/\eta$. This relationship is indeed a linear one, as the reader will immediately see from the plot.

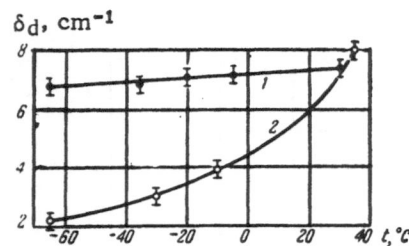

Fig. 21. Temperature variation of
the width of toluene spectral bands.
1) ν = 892 cm^{-1}; 2) ν = 1086 cm^{-1}.

Toluene. The boiling point of toluene is 110.6°C; the melting point is −95°C. The toluene molecule belongs to the point group of symmetry C_s. The problem of calculating and interpreting the frequencies of the toluene spectrum has been solved by M. A. Kovner [43]. On the basis of his calculations, we selected for our investigation the two single bands of the spectrum, the band 892 cm^{-1} corresponding to an out-of-plane fundamental vibration of the molecules in the A" symmetry class, and the band 1086 cm^{-1} corresponding to an in-plane fundamental vibration of the molecule in the A' symmetry class. The width of the IR absorption bands was measured over the temperature range from +35° to −65°C. Investigation of the band profiles revealed the shapes of the profiles to be very close to dispersion curves in both instances, except that a weak maximum, not sufficient to interfere with the determination of the width, appears on the wing of the 1086 cm^{-1} band. Graphs of the temperature dependence of the IR absorption band widths for 892 cm^{-1} and 1086 cm^{-1} are shown in Fig. 21.

As we readily see from the graph, the width of the 1086 cm^{-1} infrared absorption band narrows sharply as the temperature of the liquid rises. The reorientation potential barrier of the toluene molecule in liquid was computed from the temperature dependence of the width of this band, and was found to be U_{or} = 2.6 ± 0.5 kcal/mole. The U_{or} value so obtained was found to be in excellent agreement, within the limits of experimental precision, with the value of the potential barrier found from the temperature variation of the width of the depolarized Raman lines for the molecule, which was 2.2 ± 0.5 kcal/mole. This finding demonstrates the fact that the temperature dependence of the width of the 1086 cm^{-1} absorption band is principally due to the rotatory Brownian movement of the constituent molecules in the liquid. Further confirmation to back up this inference may be sought in the proportionality of the width of the band to the reciprocal viscosity in the case of 903 cm^{-1} band in the infrared absorption spectrum of cyclohexane.

Figure 22 presents a graph of this relationship for the 1086 cm^{-1} band, clearly evidencing the fact that the width of the IR band is a linear function of $1/\eta$.

The width of the other band investigated, that of 892 cm^{-1} (Fig. 21), varies only slightly as the temperature of the liquid is changed. The reason is probably that reorientation of the toluene molecule in the liquid is most probable about one specific axis, while the dipole moment of the vibration corresponding to that band is lined up with the direction of the reorientation axis. Actually, the selection rules dictate that the dipole moment of out-of-plane vibrations in the A" symmetry class will be lined up with the principal axis of the molecule, at right angles to the plane of the benzene ring. For that reason, the projections of the derivative of the dipole moment onto the axis of the fixed coordinates in which the observation is being carried out will not change as the molecule reorientates. It is an experimentally recorded fact that the reorientation of the toluene molecule in liquid is most probable about one of the principal axes, and this leads us to infer a certain amount of anisotropy on the part of the force field in which each molecule is immersed. In that case, the reorientation probability of the molecule about the diverse axes may be substantially different in each instance.

It is well to note that in the Raman scattering case the induced moments of the out-of-plane vibrations of the A" symmetry class of the toluene molecule do not line up with the direction of that axis, and that the width of the Raman lines corresponding to these vibrations must be a function of the temperature of the liquid. This is confirmed by the findings reported in [23], where it was shown that the width of the 217 cm^{-1} line corresponding to an out-of-plane vibration in the A" symmetry class varies appreciably in response to a change in the temperature of the molecular species.

Acetonitrile. The boiling point of acetonitrile is +82°C, the melting point is −44°C. According to data in reference [16], the solitary absorption band 918 cm^{-1} belongs to a fundamental mode of vibration. The profile of this absorption band when the substance was at room temperature has been studied [40]. That author demonstrated the profile of the 918 cm^{-1} band to have a shape close to that of a dispersion curve, but the band nevertheless exhibits pronounced asymmetry. This asymmetry is no hindrance, however, to deter-

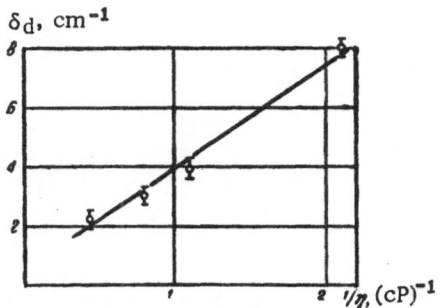

Fig. 22. Dependence of width of 1086 cm^{-1} band of toluene spectrum on reciprocal viscosity $1/\eta$.

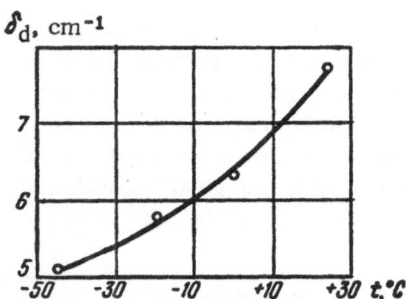

Fig. 23. Temperature dependence of the width of the 918 cm^{-1} band in the acetonitrile spectrum.

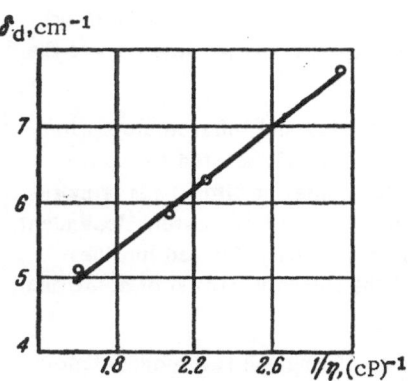

Fig. 24. Dependence of the width of the 918 cm^{-1} on the reciprocal viscosity $1/\eta$ in the acetonitrile spectrum.

mining the width of the band. The temperature variation of the width of depolarized lines in the Raman scattered spectrum of acetonitrile has not been investigated to date over a broad temperature range. We have available only A. I. Sokolovskaya's measurements [27], carried out at two different temperatures of the substance. According to these measurements, the width of the depolarized Raman lines is a sensitive function of the temperature of the liquid, and is thus indicative of the effect exerted by the rotatory Brownian motion of the molecules. It is therefore natural to anticipate the dependence of the width of the infrared absorption spectral bands of acetonitrile on the temperature of the substance.

The width of the 918 cm^{-1} absorption band was measured over a range of temperature extending from +25° to −44°C. Figure 23 shows a graph of the width of that band plotted against the temperature of the liquid. Clearly, from inspection of the graph, a drop in the temperature of the substance implies a significant narrowing of the width of that absorption band. Making use of formula (IV.2) on the temperature function of the bandwidth, we proceeded to calculate the potential barrier hindering molecular reorientation in the liquid, and found this to be 1.5 ± 0.5 kcal/mole. As mentioned earlier in Chapter II, the potential barrier to reorientation of the molecule in the liquid is the same, as a rule, as the potential barrier calculated from the temperature dependence of the viscosity. The calculations performed reveal the height of this potential barrier to be 1.3 ± 0.3 kcal/mole, which is really in fairly good agreement with the value found at 1.5 kcal/mole.

Another confirmation of the prominent role of Brownian rotatory motion of the molecules in the liquid in the temperature variation of the width of the 918 cm^{-1} band is the dependence of the width of that band on the reciprocal viscosity; this relationship will be a linear one in the case of interest. We see a graph of the width of the 918 cm^{-1} band as a function of $1/\eta$ in Fig. 24. The reader will clearly see from the graph that this is in fact a linear relation.

We infer from the experimental data cited in this chapter that random reorientations of molecules in liquids and crystalline substances exert an appreciable influence on the width of infrared absorption bands. All of the regularities in this motion of molecules such as are observed in the case of Raman scattered spectra are again observed in the infrared absorption spectra. As an example, the potential barriers hindering reorientation of molecules, as determined from the Raman spectra, agree within the limits of experimental precision with the potential barriers for the same substances as determined on the basis of the infrared absorption spectrum. A linear dependence of the width of the line (or band) on the reciprocal viscosity of the liquid is observed both in Raman scattering and in infrared absorption. Particularly noteworthy is the point that the techniques of infrared absorption and Raman scattering supplement each other in this research and in many instances smooth the way for arriving at useful inferences on the nature of the rotatory motion of molecules in the liquid and crystalline states.

SUMMARY

The experimental material presented in this article demonstrates conclusively that the investigation of the temperature relationships of the width of Raman scattering lines and infrared absorption bands will provide information on the nature of the rotatory motion of molecules in liquids and crystalline substances.

On the basis of experimental data, we carried out determinations of the mean reorientation time of a molecule τ_0 in a liquid, which was found to be $(1-10) \cdot 10^{-12}$ sec. A comparison of τ_0 and the period τ' of rocking vibrations of a molecule, $(1-2) \cdot 10^{-13}$ sec, reveals that a molecule will have time to execute anywhere from 10 to 100 vibrations between two successive reorientations. This result is in excellent harmony with currently entertained concepts on the nature of the random thermal motion of molecules in a liquid. In accordance with these concepts, the molecule executes an oscillatory motion about a "temporal" equilibrium position and in time intervals measurably longer than the period of the oscillation they will manage to effect a transition to a new equilibrium position.

In line with the theoretical concepts advanced by Ya. I. Frenkel', the mean reorientation time τ_0 must obey the relation

$$\tau_0 = \text{const} \cdot e^{\frac{U_{or}}{kT}}, \tag{1}$$

where U_{or} is the potential barrier to molecular reorientation. Our research on the relationship between the temperature and the width of Raman scattered lines and infrared absorption bands showed this relation to be validated. It is important to note here that the object of investigation in the experiments we devised was the molecule of the liquid directly.

The experimental data which we secured with the aid of the Raman scattering and infrared absorption techniques indicate that the width of the Raman lines (or IR absorption bands) is linearly related to the reciprocal viscosity of the liquid. We infer from these findings that the mean reorientation time τ_0 is proportional to the viscosity, i.e., that the Brownian rotatory motion of the molecules is to a large extent dependent upon the viscosity of the liquid. These data fit well with the data reported in papers on polarized luminescent emission, where some relationship between the viscosity of a liquid and the rotatory motion of constituent molecules of the liquid was also established.

According to the concepts put forth by Ya. I. Frenkel', we are obliged to distinguish two modes of molecular motion in a liquid: the passage of a molecule from one equilibrium position to another as a result of a change in the position of the molecule's center of gravity, and a passage to a new equilibrium position as a result of an abrupt flip in the molecule's orientation. The distinction between the rotational and translational motion resides in the fact that a molecule may rotate through large angles in a reorientation, whereas it will be displaced only over very minute distances on the order of 10^{-8} cm when the position of the center of gravity shifts. On the basis of these concepts, Ya. I. Frenkel' demonstrated the fact that the viscosity coefficient is proportional to the mean time between two displacements of the molecule's center of gravity. From that point of view, the results which we obtained are a clear indication that molecular reorientation and a displacement of the molecule's center of gravity are equiprobable events in the case of the liquids we investigated.

144

This conclusion is a further validation of the agreement we found between the potential barrier to molecular reorientation as determined by the Raman scattering and infrared absorption spectra, and the potential barrier determined from the temperature—viscosity relations.

The dependence of the reorientation potential barriers U_{cr} of the molecule from the temperature of the liquid, which we discovered, as well as the agreement between the heights of these potential barriers in both liquid and crystal, support the assumption that the height of U_{or} is a function of the relative positions of the nearest neighbors of the molecule alone. As is generally known [1], the first coordination number (i.e., the number of molecules lying on the sphere of radius equal to the radius of intermolecular coupling) either varies insignificantly when a substance effects a transition from the liquid state to the crystalline state, or else remains constant. In a liquid, the first coordination number experiences minimal fluctuations. It is therefore not at all excluded that the height of the potential barrier U_{or} of a molecule will be determined by the coupling between molecules lying on the first coordination sphere.

The results reported in the present article on investigations of the rotatory motion of molecules in liquid and crystalline substances reveal that the application of Raman scattering and infrared absorption techniques to this purpose offers opportunities of broad scope, and renders possible the determination of such important parameters as the mean reorientation time and potential barrier between reorientations of the molecule. These two techniques proved to be efficient tools in carrying out a detailed study of the nature of the rotatory motion of molecules in pure liquids, solutions, and crystalline substances. Pursuance of this line of investigation further contributed to the accumulation of a host of experimental data bearing on the nature of the rotatory Brownian movement of molecules, an especially important point in the development of the theory of the liquid state of matter, and in shedding new light on the interrelations between diverse physical properties of liquids and crystals and the motion of molecules and interactions between molecules.

The experimental data presented in the present article make it possible to correlate the available experimental evidence with the conclusions drawn in new theoretical contributions devoted to the problem of the physical nature of the width of Raman scattered spectral lines, published recently by K. A. Valiev [44-46]. This author discussed the following possible causes of broadening of the Raman lines: dissipative loss of a vibrational quantum by the molecule (conversion of vibrational energy into thermal motion), interaction between a molecule and its nearest neighbors, and the effect of reorientations of molecules. The following formula was derived to take into account the reasons alluded to for broadening, with respect to the shape of the line profile:

$$\varphi\,(\omega - \omega_0) = \left[\frac{13}{6}\frac{\rho}{1+\rho}\,\varphi\,(\gamma_1 + \gamma_2 + \gamma_3) + \frac{6-7\rho}{6\,(1+\rho)}\,\varphi\,(\gamma_1 + \gamma_2)\right], \qquad (2)$$

where

$$\varphi\,(\gamma_1 + \gamma_2 + \gamma_3) = \frac{1}{\pi}\frac{\gamma_1 + \gamma_2 + \gamma_3}{(\gamma_1 + \gamma_2 + \gamma_3)^2 + (\omega - \omega_0)^2}\;,$$

$$\varphi\,(\gamma_1 + \gamma_2) = \frac{1}{\pi}\frac{\gamma_1 + \gamma_2}{(\gamma_1 + \gamma_2)^2 + (\omega - \omega_0)^2}\;;$$

ρ is the depolarization ratio of the line, γ_1 is the width of the line due to the interaction of the molecule with its nearest neighbors, γ_2 is the "dissipative" width, $\gamma_3 = 2/\tau_0$ is the width of the line associated with molecular reorientations, and ω_0 is the frequency of the line at the intensity maximum. Formula (2) was derived by normalizing to unity the intensity at the line maximum. Further, broadening due to the interaction of a molecular vibration and rotatory motion was not taken into account in the derivation of the formula.

Formula (2) accounts excellently for all the regularities observed experimentally in the variation of the width of Raman scattered spectral lines. For example, according to the theory the width γ_1 will be independent of the temperature, whereas the width γ_3 should be a sensitive function of the temperature:

$$\gamma_3 = \text{const}\cdot e^{-\frac{U_{or}}{kT}}\,. \qquad (3)$$

Now, experiment reveals γ_2 to be only slightly temperature-dependent, and so this contribution may be neglected.

Consider the consequences flowing from formula (2). In the $\rho = 0$ case, the width of the line will be $\gamma_1 + \gamma_2$ and will be independent of the rotatory motion of the molecules, whereas at $\rho = \frac{6}{7}$ the width of the line will be determined by the sum $[\gamma_1 + \gamma_2] + \gamma_3$. The first two terms are practically independent of the temperature, but γ_3 obeys the temperature law (3). The shape of the line profiles of width $\gamma_1 + \gamma_2$ and $\gamma_1 + \gamma_2 + \gamma_3$ displays that of a dispersion type curve. It is precisely these regularities which we happened to observe in the experiment. Thus, from formula (2) for lines such that $\rho \approx \frac{6}{7}$, we arrive at the empirically derived relationship

$$\delta\,(T) = \delta_0 + Ae^{-\frac{U_{\mathrm{or}}}{\kappa T}}, \tag{4}$$

where $\delta_0 = \gamma_1 + \gamma_2$ is the temperature-independent "residual width."

Clearly, from formula (2), lines such that $\rho \approx \frac{6}{7}$ have a greater width than do lines of low ρ ratio. In the transition to more complex molecules, of low reorientation probability in a liquid, the value of γ_3 tends to zero, and we then infer from (2) that the width of the depolarized Raman lines will be equal to the width of Raman lines of low depolarization ratio. This was precisely the case observed by N. I. Rezaev in the case of the spectral lines of N-butylbenzene and N-hexylbenzene [18].

All of the changes in line width observed experimentally are accordingly found to be in excellent harmony with the theory expounded by K. A. Valiev.

In conclusion, the author welcomes the opportunity to express his grateful gratitude to Professor Pavel Alekseevich Bazhulin, in charge of research, for his persistent attention and kind assistance in the completion of the above research project.

LITERATURE CITED

1. I. Z. Fisher, Statistical Theory of Liquids, Fizmatgiz (1961).
2. M. Leontovitch, J. of Phys. [USSR] 4:499 (1941).
3. I. I. Sobel'man, Izv. Akad. Nauk SSSR, Ser. Fiz. 17:554 (1953).
4. Ya. I. Frenkel', Selected Writings, Vol. III, Izd. AN SSSR (1950).
5. N. Bloembergen, E. M. Purcell, and R. V. Pound, Phys. Rev. 73:679 (1958).
6. M. V. Aleksandrov, Usp. Khim. 29(9):1138 (1960).
7. I. L. Fabelinskii, Tr. Fiz. Inst. Akad. Nauk 9:183 (1958).
8. E. Gross, Nature 126:201, 603 (1930).
9. E. Gross and M. Vuks, J. Phys. Radium 6:457 (1935).
10. M. F. Vuks and V. L. Litvinov, Dokl. Akad. Nauk SSSR 105:696 (1955); A. K. Atakhodzhaev, M. F. Vuks, and V. L. Litvinov, Proceedings of the Tenth Conference on Spectroscopy, Vol. I, L'vov, p. 118.
11. S. I. Vavilov and V. L. Levshin, Z. Physik 16:135 (1923).
12. V. L. Levshin, Liquid and Solid-State Photoluminescence, GITTL (1951).
13. A. N. Sevchenko, Tr. Gos. Optich. Inst. 14:65 (1941).
14. V. I. Gribkov and N. D. Zhevandrov, Dokl. Akad. Nauk SSSR, 98:565 (1954).
15. L. Landau and G. Lifshits, Statistical Physics, GTTI (1951); English translation, Addison-Wesley, Cambridge, Mass.
16. G. Herzberg, Vibrational and Rotational Spectra of Polyatomic Molecules (Russian translation of "Infrared and Raman Spectra of Polyatomic Molecules," Van Nostrand, 1945), IL (1949).
17. M. L. Sosinskii, Izv. Akad. Nauk SSSR, Ser. Fiz. 17:621 (1953).
18. N. I. Rezaev, Vestn. Mosk. Univ. (2):145 (1957).
19. Rayleigh, Phil. Mag. 42:441 (1871).
20. S. G. Rautian, Usp. Fiz. Nauk 66(3):475 (1958).
21. N. I. Rezaev, Proceedings of the Tenth Conference on Spectroscopy, Vol. I, L'vov (1957), p. 230.
22. A. V. Rakov, Opt. i Spektroskopiya 7:202 (1959).
23. P. A. Bazhulin and A. I. Sokolovskaya, Research in Experimental and Theoretical Physics. Memorial Volume — G. S. Landsberg, Izd. AN SSSR (1959), p. 56.
24. G. V. Mikhailov, Proceedings of the Tenth Conference on Spectroscopy, Vol. I, L'vov (1957), p. 227.
25. Chemical Handbook, Vol. I, Goskhimizdat (1951).
26. Power Engineering Handbook, Vol. 10, Sovetskaya Entsiklopediya (1930).
27. A. I. Sokolovskaya, This issue, p. 63.
28. H. Eyring, Theory of Absolute Reaction Rates [Russian translation of: S. Glasstone, K. Laidler, H. Eyring, Theory of Rate Processes, McGraw-Hill, 1941], IL (1948).
29. S. D. Ravikovich, Structure and Physical Properties of the Liquid State, Izd. Kiev State Univ. (1954), p. 87.
30. M. F. Vuks, Opt. i Spektroskopiya 9:92 (1960).
31. E. Andrew, Nuclear Magnetic Resonance [Russian translation], IL (1957); original: Cambridge Univ. Press (1955).
32. A. S. Davydov, Theory of Light Absorption in Molecular Crystals, Kiev, Izd. AN Ukr. SSR (1951).
33. M. A. Kovner, Dokl. Akad. Nauk SSSR 97:229 (1954).
34. C. R. Bailey, C. K. Ingold, and S. C. Carson, J. Chem. Soc., (April):252 (1946).

35. M. M. Sushchinskii, Tr. Fiz. Inst. Akad. Nauk 12:54 (1960).

36. E. R. Andrew and R. G. Eades, Proc. Roy. Soc. A216:398 (1953).

37. M. N. Markov, Dokl. Akad. Nauk SSSR 108:428 (1956).

38. R. N. Tones et al., Rev. Universelle Mines, 9e Serie, 15(5):417 (1959).

39. A. V. Rakov, Opt. i Spektroskopiya 13:369 (1962).

40. G. G. Petrash, Opt. i Spektroskopiya 9:121 (1960).

41. V. L. Malyshev and S. G. Rautian, Izv. Akad. Nauk SSSR, Ser. Fiz. 23:1237 (1959).

42. M. P. Lisitsa et al., Opt. i Spektroskopiya 8:638 (1959).

43. M. A. Kovner, Dokl. Akad. Nauk SSSR 97:65 (1954).

44. K. A. Valiev, Zh. Eksperim. i Teor. Fiz. 40:1832 (1961).

45. K. A. Valiev, Opt. i Spektroskopiya 11:465 (1961).

46. K. A. Valiev and L. D. Eskin, Opt. i Spektroskopiya 12:758 (1962).

INVESTIGATION OF THE STRUCTURE AND WIDTH
OF RAMAN SCATTERING LINES IN GASES
UNDER HIGH PRESSURE

G. V. Mikhailov

CHAPTER I

COLLISION BROADENING
OF MOLECULAR SPECTRUM LINES
IN GASES

§1. Theory of Collision Broadening and Shift of Spectral Lines in Gases

In a theoretical investigation of intermolecular interactions, the force which is due to the interaction of individual particles is usually separated out. This is the basis of calculations which consider one or another interaction. In specific calculations, the nature of the particle motion in the surrounding medium is taken into account.

A substantial number of articles have been devoted to the study of intermolecular forces. In view of the complex nature of intermolecular forces, calculations of them are of a qualitative character. They are usually expressed by semiempirical formulas.

A specific calculation of the intermolecular interaction for several simple cases has been given in [1].

The perturbation experience by a molecule depends fundamentally on the medium in which it is found. According to the character of the perturbation, media can be conveniently divided into two groups: rarefied gases, and condensed media (gases at high density, liquids, and solids).

The interaction mechanism is simplest in the case of a rarefied gas. The molecule experiences infrequent collisions, and remains unperturbed the greater part of the time. The number of effective collisions is determined by particle density, particle speed, and the character of the interaction during the individual collisions, which depends upon the forces which act between the particles and the parameters of the collision: the relative velocity, the impact parameter, and the orientations of the colliding particles. Quantitatively, the result of the interaction is expressed by an effective collision diameter.

In condensed media the interaction is of a more complex nature, since a molecule is constantly in the perturbing field of its surroundings. Determination of this field requires an understanding of the forces and the character of the particle motion in the medium. Such data are absent in practice.

In rarefied gases, as in condensed media, the interaction of a molecule leads to a change in intensity, a frequency shift, and a broadening of the radiated lines. Among these changes, the most fully studied are the broadening and shift of the spectral lines. A survey of the experimental and theoretical work on broadening and shift up to the beginning of the present investigation is given in [1] and [2].

Theories of broadening have been developed for a gas decaying on collision, which describes broadening in rarefied gases, and also a statistical theory for the case where the perturbation can be put in the form of a field which changes slowly in comparison with the radiation frequency. The detailed theory of broadening is found in [2] and [3].

_____

*Dissertation in partial fulfillment of the requirements for the degree of Candidate in Physicomathematical Science. Defended October 24, 1962, before the Physics Faculty of M. V. Lomonosov Moscow State University. Research Director: Doctor of Physicomathematical Science Prof. P. A. Bazhulin.

Below is a short account of certain problems of the theory necessary for the present work.

At the basis of the Lorentz collision broadening theory lies the idea of infrequent collisions with a sharp change in the radiation at the time of collision; i.e., a change of phase or a break of the radiation. The radiation is broken into a series of independent trains, which leads to a broadening of the line. The broadened line has a Lorentzian shape. Its width 2γ, in the case of identical particles, is given by the expression

$$2\gamma = 2N v_{rel} \sigma, \tag{L1}$$

where N is the number of particles per unit volume, v_{rel} is the relative velocity, and σ is the effective optical collision cross section.

The quantity σ shows which collisions can be considered effective in the scattering mechanism. To determine σ, Lorentz and Weisskopf consider as effective those collisions in which the correlation between oscillations before and after the collision is destroyed. This can be estimated quantitatively through the change η in the phase of the radiation, which arises due to the change of frequency at the time of the collision. According to Weisskopf, encounters which lead to a radiation phase shift $\eta > 1$ can be considered effective. This makes it possible to determine the largest value of the impact parameter ρ at which encounters are still effective (the so-called Weisskopf radius ρ_0) and the effective optical cross section σ which is related to ρ_0 by the expression $\sigma = \pi \rho_0^2$. The limiting value of ρ_0 (determined by $\eta_0 = 1$) was chosen arbitrarily by Weisskopf. Some of the arbitrariness in the choice of η_0 is removed in calculations of the impact broadening by means of correlation theory. Actually, these calculations show that encounters at impact parameters greater than some ρ_0 do not contribute substantial broadening; thus the calculated limiting value of ρ agrees with the Weisskopf radius ρ_0 considered above.

If the character of the interaction which determines the frequency shift of the radiation at the time of collision is known, the theory enables one to obtain an analytical expression for σ. It should be noted that the calculation of σ for a molecule is extremely complex, and usually is of an approximate character.

Radiation at the moment of collision is not taken into account in the collision theory. This idea is justified as long as the collision time is much less than the mean free time, i.e., as long as the average distance to a perturbing particle $\overline{R} = (3/4\pi N)^{1/3}$ is much greater than the Weisskopf radius ρ_0. This gives an upper limit to the pressure at which the collision theory applies:

$$\rho_0^3 N \ll 1. \tag{L2}$$

In the statistical theory of broadening, we consider the radiation of a particle in the fluctuating field of its surroundings; the changing field is considered small in comparison with the radiation frequency ω_0. The shift of the level (and consequently of the radiation frequency) in this field leads to a broadening of the lines. If the nature of the interaction is known, the statistical theory allows one to find the broadening and shift of the lines. Such a calculation for a simple interaction of the Van der Waals type $(1/R)_6$, gives a broad line and a frequency shift.

At small densities (where collision broadening occurs) the statistical theory describes the interaction at the moment of collision. In this case the perturbing field is due to encounters inside the Weisskopf radius ρ_0; this perturbation causes the distribution of intensity in the far wings of the line. The general contribution of this broadening to the integrated line intensity depends on the ratio of the collision time to the mean free time. In the region of collision broadening, this contribution is small, and the most important part is given by the previously considered collision mechanism, which determines the broadening in the center of the line.

The theory described relates to the broadening of separate isolated lines in molecular bands. With a substantial broadening, level overlap can begin, which leads to a redistribution of energy in the spectrum. Thus in the case of rotation spectra, a transfer of energy into the central band corresponding to the transition $\Delta J = 0$ can be observed [3].

Despite collision broadening, a substantial contribution to the observed line breadth in molecular spectra can be due to structure, and its role in the breadth of a line will be specially considered in Chapter 4.

§2. Experimental Data on the Broadening of Molecular Spectrum Lines in the Gas Phase

A. Broadening in Molecular Infrared Absorption Bands. The general laws of behavior for the broadening of the rotational components in a rotation-vibration absorption band under the action of pressure have been investigated by Kortum and Verleger [4] in the HCN molecule. They studied the broadening of the rotational lines in the P and R branches of the rotation-vibration band at 1.0380 μ (vibrational transition 0-3) under the action of the intrinsic HCN gas pressure and impurity gases in the pressure range from 75 to 550 torr.

For a broadening by the intrinsic pressure of the gas, the breadth of the rotational components of the P and R bands is greatest for the values of J which correspond to the most populated levels, in which the dependence of the breadth on J is suggestive of the thermal distribution of the intensity in a rotational band. For the HCN molecule, the dipole and resonant rotational interactions can be observed. A calculation of the broadening produced by dipole —dipole interaction leads to the conclusion that the greatest breadth can be expected for small values of J. The resonant interaction of identical molecules leads to a broadening which is greatest for the most populated levels. The dependence of the breadth on J in this case is analogous to the thermal distribution of intensity in a rotational band. Both types of interaction are known to decrease the broadening for large J values [5-7].

Inasmuch as the experimental dependence of the breadth on J resembles thermal broadening, comparison with calculations indicates that in broadening by the intrinsic pressure of the gas the most important part of the broadening in the rotational components of the 0-3 vibrational band of HCN is due to the resonant interaction. The absolute values of the breadths of the rotational components are less for the R branch than for the P branch; this difference has been predicted theoretically by Lindholm [8]. In the pressure range investigated, the average values of the breadth of the rotational components, for a given density, are directly proportional to the pressure. This agrees with the fundamental assumption of the Lorentz broadening theory [2].

For impurity gas broadening of the rotational components of the P and R branches of the 0-3 HCN band, experiment shows that the greatest breadth corresponds to levels with small values of J, i.e., the breadth decreases with increasing J. In the given case, the resonance interaction is excluded, but the dipole interaction leads to a dependence analogous to that observed experimentally. Studies of the collision breadth of HCN lines in the presence of impurity gases show that the broadening increases in direct proportion to the dipole moment of the perturbing molecule.

Investigations of the broadening of the rotational components in the near infrared absorption bands of HCl, H_2O, and CO_2 [4] show that the experimental data are in agreement with the Lorentz theory; the lines have the Lorentzian shape and the half-width increases in proportion to the pressure.

Weber and Penner [9] investigated the broadening of rotational lines in the infrared spectra of NO, HCl, and HBr. H_2, He, and Ar were used as perturbing gases. The effective optical cross section for broadening which they obtained was somewhat less than the gas-kinetic cross section. The experimental results of these authors fit into the general theoretical picture of broadening.

An investigation of the broadening created by various impurity gases has been carried out by Coulon et al. [10]. They investigated the perturbation of the fundamental HCl band (3000 cm^{-1}) by the gases N_2, Ar, He, H_2, and O_2, and of the 2100 cm^{-1} band of CO by the gases N_2 and Ar, with pressures up to 1000 atm. It was found that the perturbation increases in the following order: He, H_2, Ar, O_2, N_2. For the HCl and CO molecules, a linear increase of absorption intensity with increasing pressure is observed, corresponding to the increase of density with increased pressure. In the HCl spectrum with a N_2 partial pressure, there is a new line near the forbidden Q branch. The intensity of this line increases rapidly with pressure.

The effect of an impurity gas has been studied by Benesh and Elder [11], who investigated the broadening of the separate rotational lines J = 2 → J = 3 in the 3.4 μ absorption band of HCl and J = 4 → J = 5 in the 3.25 μ band of CH_4. They calculated relative values of the effective optical collision diameter for HCl and CH_4. This comparison allowed them to divide gases into two groups according to the character of the perturbation they introduce. One of these groups corresponds to molecules with high symmetry and the other to dipole molecules; the established connection between the dipole moment and the amount of broadening was not observed by these authors.

In addition, they carried out an analogous comparison of impurity gas broadening for HCl and NH_3. For NH_3 the literature data on broadening of the absorption lines in the microwave spectrum was used [12, 13]. The comparison shows that in spite of the very different frequency regions and the different character of the molecules, the same broadening mechanism acts in all cases.

B. Line Broadening in the Microwave Region. Broadening under the action of intrinsic gas pressure has been investigated in the microwave region for molecules of ammonia [13-15], oxygen [16], water [17, 18], metal halides [19], other molecules such as OCS [14, 20, 21], and cyanogen halides [22, 23]. At low pressures, the line breadth increases linearly with pressure and no shift is observed. For all of the molecules indicated above, the effective collision radius ρ_0 has been found.

The oxygen molecule, whose absorption is related to its magnetic moment, has been investigated in [24] and [25]. The average broadening of the lines in the absorption band amounts to 0.0475 cm^{-1}/atm. Hill and Gordy [1] have measured the width of the 2.15 mm absorption lines in O_2 in the microwave region at 300°K and 195°K, and obtained 0.064 and 0.09 cm^{-1}/atm broadening, respectively. With a change of temperature, as is well known, the number of collisions changes, and also the character of the interaction during the individual collision, i.e., the time and energy of interaction. The possibility of using data on the temperature dependence of the broadening to study the character of the interaction is discussed in [26]; the temperature dependence of the broadening for NH_3 and O_2 has been calculated by Mizushima [6].

Experimental and theoretical investigations of the dependence of the line breadth on the quantum number K in the oxygen spectrum (just as on J in the infrared spectrum) [27, 24, 28] reveal the presence of a resonant rotational interaction, which makes a substantial contribution to the broadening of the microwave spectral lines of oxygen in addition to the contributions from quadrupole and other types of interaction.

Broadening of microwave spectral lines by impurity gases has been investigated in the spectra of NH_3 [11-13, 29], H_2O [18], and O_2 [24, 30]. Measurements of the effective cross section show that the greatest broadening is produced by dipolar gases; the absolute value of the cross section is, on the average, of the same order as the gas-kinetic cross section. In the case of strong long-range interactions, the effective optical cross section is larger than the gas-kinetic cross section.

C. Line Broadening in the Raman Scattering Spectrum. In the area of Raman scattering, the experimental material on broadening and shift of lines is extremely meager. There are no investigations of the dependence of the line breadth on pressure in gases.

Welsh and his collaborators [31] have studied the Raman scattering spectrum of CO_2 at pressures from 15 to 220 atm in the gas and the liquid. The investigation of the breadth in this work had a qualitative character: the rotational bands at the indicated pressures were completely overlapping, and only changes in the shape of the band as a whole were considered. The authors observed a broadening of the rotational wings of the vibrational band with increasing pressure, and in the transformation into the liquid state. However, for the Q branch of the vibrational band of CO_2 they did not discover any observable broadening, even with a transition into the liquid state.

It is known that a vibrational line in Raman scattering has a substantial breadth, the origin of which is insufficiently clear. Kh. E. Sterin [32, 33] investigated the behavior of the line breadth of the fully symmetric vibration of benzene ($\nu = 992$ cm^{-1}) during the transformation from vapor to liquid, and discovered that it does not change. I. L. Sobel'man [3, 34] has stated the hypothesis that the substantial broadening and the fact that it is preserved during the phase transition is connected with the structure of lines which arise from the interaction of vibration and rotation. His approximate estimate demonstrates the possibility of such an explanation.

In condensed media, the character of the intermolecular interaction is considerably more complex; here one can expect both broadening and also a change (splitting) in the structure of the lines [35].

Welsh et al. [36, 31] investigated the Raman scattering spectra of O_2, N_2, CO_2, and CH_4 in the liquid state and observed rotational wings on the vibrational bands of these molecules. However, the considerable broadening of these bands did not allow a study of their structure. The Q branch of the vibrational bands of O_2, N_2, and CO_2 remained sharp in the liquid.

The survey given shows that until the beginning of the present work, there was essentially no systematic investigation of the nature of the line breadth in Raman scattering in gases. Recently several papers have been published which are not mentioned in the survey. These will be considered during the discussion of our results.

It should be noted that, in contrast to infrared absorption, Raman scattering in the microwave region also allows the inclusion of a broad group of nondipole molecules and the study of their interactions. In the absorption spectra of molecules, only those transitions which continue to appear in Raman scattering are usually observed. The method allows inclusion of a significantly wider frequency range and permits full study of the interaction processes.

In considering the nature of the line breadth, the most promising investigation is to study the basic parameters of lines in gases during a change of pressure, where the mechanism of molecular motion is simplest and the theory of broadening has been worked out. The present work is devoted to an investigation of the structure and collision broadening of the Raman scattering lines of simple molecules, in a gas, under the action of intrinsic pressure.

CHAPTER II

EXPERIMENTAL APPARATUS

§1. Introduction

The basic arrangement of equipment for studying Raman scattering is shown in Fig. 1. The substance in the illuminator O is illuminated by monochromatic light, and the light scattered by the substance is transmitted by the condensing system K into the spectrograph Sp, where it is decomposed into a spectrum and recorded by the output detector P.

For a given spectral apparatus with a properly chosen condensing system, the intensity of the scattered light entering the spectrograph is given by the expression [37]

$$\mathscr{I} \sim \mathscr{I}_0 \omega^4 N l. \tag{II.1}$$

Here $\mathscr{I}_0$ is the intensity of the exciting radiation, determined by the output of the lamp and the construction of the illuminator; ω is the frequency of the exciting radiation, N is the density of scattering molecules, and l is the length of the scattering volume, usually corresponding to the length of the discharge arc of the lamp. The effective length used can be increased by means of a mirror. Full utilization of the scattered radiation is achieved by use of a specially designed condensing system, a large aperture spectrograph and a sensitive detector.

In our case the problem of constructing the apparatus was solved by developing an illuminator for studying scattering under pressure, and by effective utilization of the scattered radiation by means of mirrors. A general view of the experimental apparatus is shown in Fig. 2.

§2. The Source of Excitation

The mercury discharge lines Hg 4047 and Hg 4358 A were used to excite the scattering spectrum. PRK-2 mercury lamps were used as sources for these lines in the high-pressure illuminator. In the illuminator they were placed inside a special cooling cylinder of molybdenum glass. The cylinder was formed from two jackets; the inside part of it was cooled by circulating water, and the outside filled with a liquid light filter. Outside the cylinder a foil was wrapped. The space between the glass lamp and the cooling cylinder was about 2 mm. The lamps, thus cooled, operated with a somewhat diminished mercury pressure.

In choosing the operating regime, the dependence of intensity and shape of the exciting line contour on lamp voltage was mapped. The best operating regime was I = 3.5 A and V = 70 V. Under these conditions there was no self-reversal and the excitation line was still sufficiently intense. Both lines (4047 and 4358 A) had a breadth $2\gamma \approx 1.5$ cm^{-1}.

The continuous background in the exciting spectrum was eliminated during mapping of vibration lines by a liquid light filter (0.1% I in CCl$_4$) with an absorbing layer 5 mm thick. The filter was changed after 40 to 50 hr of work. In recording a pure rotational band ($\Delta v = 0$) the filter was not used.

For investigations in liquid nitrogen, lamps of the Sosinskii type with water-cooled electrodes were used in the illuminator [38]. Working in the regime 10 A and 30 V, these lamps gave a narrow excitation line

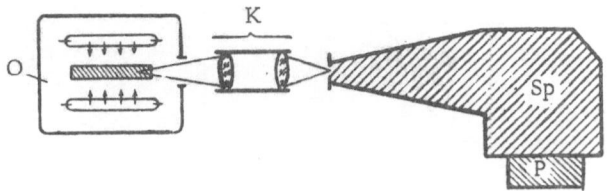

Fig. 1. Basic arrangement of apparatus for studying
Raman scattering.

Fig. 2. General view of the apparatus for studying Raman scattering at high
pressure. 1) 400 atm illuminator; 2) condenser system; 3) HUET B-III spectrograph; 4) TC-15 ultrathermostat.

(for Hg 4047 and Hg 4358 A, $2\gamma \sim 0.3$ cm^{-1}) with an excitation line intensity approximately five times less
than with the PRK-2 lamps. There was practically no continuous background in the spectrum.

§3. Illuminator for Investigation of Raman Scattering in Gases at Pressures up to 400 Atm

The high-pressure illuminator for investigations of Raman scattering in gases constructionally combined
a system for illumination and a high-pressure scattering vessel. Sealed transparent windows of considerable
size were designed, for pressures of several hundred atmospheres, presenting a complex construction problem.
Flat windows described in the literature, which clamp from outside [31], did not solve the problem posed: the
size of the window is small and the limit of allowable pressure is low. In our design, a plastic cylinder serves
as an illuminating window, the sealing of which is accomplished by "heat fitting." The window is simultaneously a cylindrical lens, which focuses the discharge arc of the lamp onto the scattering volume [39].

A diagram of the illuminator is shown in Fig. 3. The substance under investigation is illuminated by
three PRK-2 lamps, placed symmetrically 120° apart, parallel to the scattering volume. The light passes

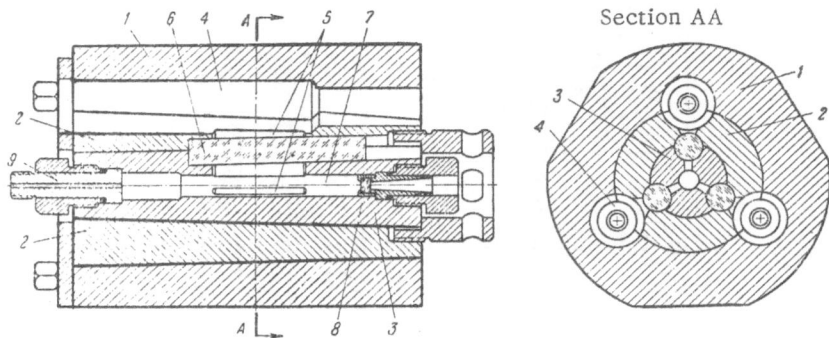

Fig. 3. Drawing of the illuminator for high pressures to 400 atm. 1) External cone body; 2) middle cone; 3) inside cone; 4) lamp recess; 5) illuminating windows; 6) cylindrical lens; 7) scattering volume; 8) observation window; 9) plug for admission of gas.

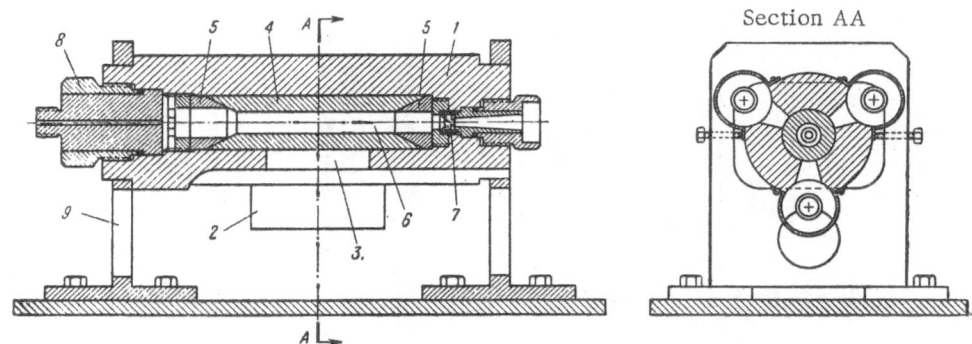

Fig. 4. Diagram of the illuminator for pressures to 1500 atm. 1) Body; 2) recess for lamp with cooling filter; 3) illuminating window; 4) cylindrical sealing window; 5) tightening cones; 6) scattering volume; 7) observation window; 8) plug for admitting gas; 9) stand.

through the scattering volume and is reflected by an aluminized mirror on the side of it, such that each ray passes twice through the scattering volume.

The body of the illuminator is made from chrome alloy steel and is composed of three cones which fit into each other and are pressed from the end to create preliminary stress. The scattering volume is located inside the inner cone and has the form of a cylinder of diameter 24 mm and length 250 mm. In one end of the scattering chamber there is an observation window. The seal between the observation window and the supporting plug is made by the fitting of the observation window in optical contact with the supporting surface of the plug. The other end of the scattering volume is closed by a plug with a tube for admitting gas. To seal the plugs, packings of Teflon, lead, and copper are used. In the side walls of the scattering volume three symmetrically placed illumination windows, of dimensions 10×100 mm, are cut. Inside the scattering volume there is a multiple-reflection mirror system. There is a hole for the cylindrical lenses between the inside and middle cones of the body. The precisely fabricated plastic cylinders have a diameter of 0.3 mm, greater than the diameter of the hole. During assembly, the cylinders are cooled in liquid nitrogen and are placed in the holes in the frozen state. Expanding, they create the necessary seal. The PRK-2 lamps and their cooling filters are placed in recesses between the middle and outer cones of the body. The whole illuminator is mounted on a special plate which is used to produce precise alignment. With regard to efficiency, without the mirror system the illuminator provided twice the intensity of the standard elliptical illuminator with one lamp, usually employed in Raman scattering.

To test the strength of the illuminator, pressure up to 450 atm was applied, after which a flow of gas was observed, caused by squeezing out of an illuminating window. With a drop in pressure the flow ceased.

During operation, the gas in the illuminator was supplied directly from cylinders. The pressure was controlled by a manometer.

Coworkers E. F. Shcherbakova and L. E. Surkov of the Institute for Very High Pressure Physics, Academy of Sciences, USSR, took part in the construction and fabrication of the illuminator. The author acknowledges his debt and expresses his appreciation for this assistance.

§ 4. Illuminator for Raman Scattering Investigations in Gases at Pressures up to 1500 Atm

A similar scheme of symmetrical illumination by three lamps was used in the illuminator for 1500 atm. A constructional peculiarity of the illuminator is the sealing illumination window: a hollow cylinder of plastic which operates on the principle of self-sealing. The use of this seal allowed us to raise the working pressure range to 1500 atm. A diagram of the illuminator is shown in Fig. 4. The body of the illuminator and the mounting were made from 40 X steel. Before polishing they were hardened, with subsequent lowering to a Rockwell hardness $R_C = 35-40$. The inside of the body of the illuminator was a precisely polished cylindrical cavity of diameter 60 mm. An observation window was placed in the front of the cavity and at the back there was a plug for admitting the gas; these pieces are the same as in the previously described illuminator for 400 atm. Three symmetrically placed illuminating windows of dimensions 10×100 mm were cut in the sides of the body.

The sealing of the illuminating windows was achieved with a complete plastic cylinder placed inside the cavity. The diameter of the cylinder exceeded the diameter of the cavity by 0.5 mm. During assembly the cylinder was cooled in liquid nitrogen, placed inside the body while cold, and tightened with the end cones. The expansion during the subsequent warming up provided the necessary preliminary sealing of the windows. With a high pressure in the illuminator, the cylinder is pressed out to the illuminating windows by the gas inside. This makes the seal at high pressures.

The scattering volume is in a cavity in the plastic cylinder, which has a diameter of 24 mm and a length of 250 mm. It allowed the same mirror system to be used as in the previous illuminator. The scattering volume was illuminated by three PRK-2 lamps, arranged as in the previous scheme. The light passes through windows in the body and through the sealing cylinder. In contrast to the previous scheme, the focusing of the discharge arcs of the lamps onto the scattering volume was eliminated in this scheme; consequently the intensity of illumination in the scattering volume (and the efficiency of the whole illuminator) was approximately three times less than previously.

The illuminator was strength-tested to the design pressure, after which it was used in the investigation of Raman scattering in methane. When the illuminator was in use at high pressure, it was surrounded by a steel shield 4 mm thick for safety.

§ 5. Multiple-Reflection Mirror System

To increase the effective use of the scattered radiation we used a multiple-reflection system of four mirrors [40]. It consisted of spherical mirrors of identical curvature (Fig. 5). The two front mirrors B and D make up a whole mirror, with a slit in the middle for passage of light; its center of curvature was located on the optical axis of the system at point O (between mirrors A and C). The back mirrors A and C were slightly inclined with respect to each other, so that their centers of curvature were located on the edge of the entrance slit of the front mirrors at points O_1 and O_2. Consequently, the aligned system—the image of the exit slit created by the mirrors cover all of the front mirror within the regions h_1, h_2, etc. — and the light rays from an elementary volume inside the system are incident on the exit slit after a certain number of reflections and fill the collimator D' of the whole area of both the front and the back mirror is used. Theoretically, the four-mirror system allows a gain of intensity by a factor of $1/(1-r)$, where r is the reflection coefficient of the mirrors.

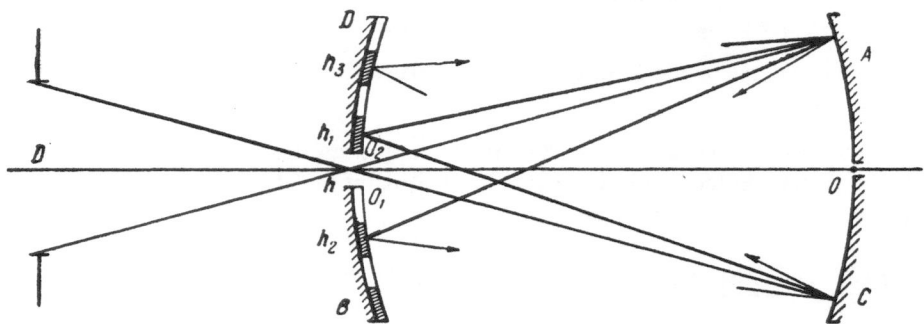

Fig. 5. Scheme of the four-mirror multiple-reflection system.

The mirror system is mounted inside the scattering volume of the high-pressure illuminator. In our illuminator, the diameter of the scattering volume is 25 mm. This allows us to secure the necessary strength of the illuminator and is sufficient for installation of the special mirror system. In the scattering volume, the mirror system (considering the diameter of its body) could have a front mirror diameter not more than 20 mm. For the back mirror, the construction of the illuminator allowed a somewhat larger diameter, up to 30 mm.

The diameter of the mirror is directly related to the width of the exit slit and the attenuation of a light beam during multiple reflection in the system. A light beam falling on the mirror at a distance d/2 from the slit undergoes m = d/h reflections before falling onto the slit, and hence is attenuated by a factor r^m. In order that the contribution from the edges of the mirror is not less than 0.05 (for an aluminzed mirror with r = 0.9), the width of the exit slit of a mirror system with d ≈ 20 mm should not exceed h ≤ d/m, i.e., 0.7 mm. In our case the slit in the mirror system was 0.4 mm wide. The image of the slit in the mirror system, created by the condensing system, should overlap by a factor of two or three the maximum possible working slit of the spectrograph. If the maximum width of the spectrograph slit is limited to 20 cm^{-1}, then the width of the image of the mirror system slit h' should be not less than 0.3 mm. In this case the image of the slit at the spectrograph should be produced without magnification: h/h' ≈ 1. For photographic recording a slit of several millimeters length is sufficient. With an image ratio of 1 : 1 the length of the slit in the mirror system should be the same. In the mirror system, the length of the slit was chosen to be S = 8 mm.

The distance l between mirrors is determined by the length l_0 of the discharge arc of the lamp, the dimensions of the collecter, the shielding of the mirrors from direct light of the lamps, and the parasitic light scattered on the walls of the system. In our case (l_0 = 100 mm) the distance between mirrors was chosen to be l = 200 mm.

Returning to the design of the back mirror, in Section 7 of this chapter it will be shown that the general formulas for the design of a scattering vessel, given in [37], apply to this mirror system. With the known values of the focal length and diameter of the collimator objective, and the chosen values of l and h/h', it can be shown that the back mirror should have a diameter D = 22 mm. To over-fill the collimator, its diameter was taken equal to 24 mm.

Thus the basic dimensions of the mirror system are the following: slit — 0.4 × 8.0 mm, d = 20 mm, D = 24 mm, and l = 200 mm.

The construction of the mirror system is shown in Fig. 6. The mirror system was mounted in a brass tube with three illuminating windows in the central part; the regions between the illuminating windows were aluminized. To eliminate parasitic light in the tube, diaphragms, consisting of blackened grooved tubes, were placed before the front and back mirrors. The mirrors were made of glass, with an evaporated aluminum coating (r = 85%). The front mirror was composed of two pieces cemented to a glass ring. In the center of the mirror there was a 0.4 × 8.0 mm slit. The mirror was placed directly on a bearing surface in the frame, precisely fabricated with a hollow in it. The back mirror was composed of two parts, cemented to a special brass holder. The holder was pressed down by a spring to a supporting plate. The back mirror assembly was placed inside a sleeve which screwed onto the body tube, thus providing lengthwise alignment of the mirror. The

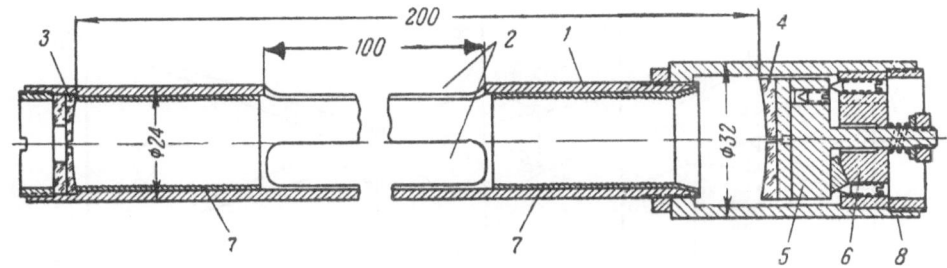

Fig. 6. Construction of the multiple-reflection mirror system. 1) Body; 2) windows; 3) front mirror; 4) back mirror; 5) holder for the back mirror; 6) bearing plate; 7) grooved tubes; 8) sleeve.

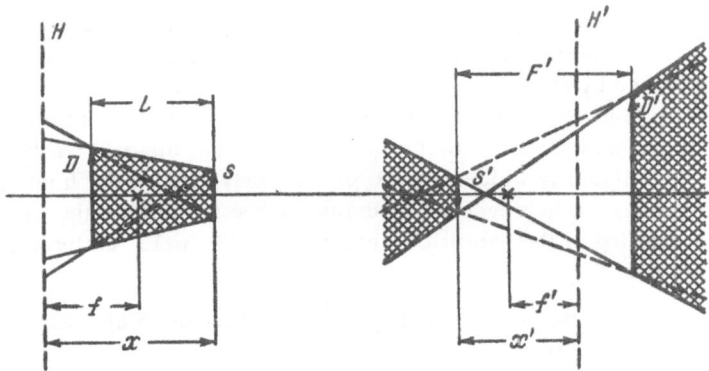

Fig. 7. System for imaging the source volume at the slit.

construction of the mirror mounting permitted a combined rotation of both mirrors, or slanting of one mirror with respect to the other. This adjustment was accomplished by means of a taper on the filling plug which fixed the rotation of the edge of the mirror with a precision of two microns and did not break with shock. The aligned mirror system was placed in the scattering volume of the illuminator and clamped by the taper. The efficiency of the mirror system was tested on a model. Photographs of the Raman scattering in benzene showed that the mirror system used gave a sixfold increase in intensity, which agrees with theory for aluminzed mirrors with r = 0.85. The level of parasitic light in the illuminator with the mirror system was the same as in standard illuminators with the usual scattering vessels.

§6. Condensing System for Imaging the Source Volume

For maximum use of the radiation from the source volume, S. G. Rautian [37] has proposed to produce an image of the radiating volume in the collimating region of the spectral apparatus, so that the front side of the vessel is imaged on the objective of the collimator and the back side on the slit of the spectrograph. This allows uniform filling of the collimator and uniform illumination of the spectrograph slit; and also allows a more rational method of calculating the dimensions of the scattering vessel. The particular system of images proposed by Rautian was achieved by a single-lens condenser.

The dimensions of the source volume were determined by the dimensions of the slit in the front mirror, the distance between mirrors, and the diameter of the back mirror. Proceeding from the shape of the source, we employed a condenser system which produced an image such that the slit in the mirror system (front wall of the vessel) was imaged on the slit of the spectrograph and the back mirror on the objective of the collimator, i.e., the image was opposite to that suggested in [37]. Such a system is specifically produced by a two-lens condenser. *

---

*Description of the application of a two-lens condenser to the image of a mirror system can be found in the literature [41]. However, there is no treatment or calculation of the system of images.

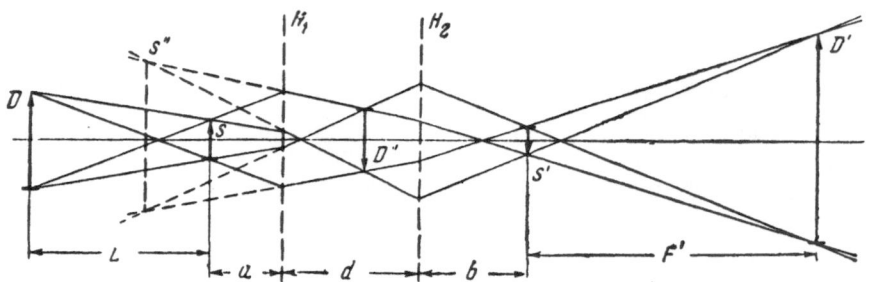

Fig. 8. Ray paths in the two-lens condenser.

The system of images is shown in Fig. 7. Here H and H' are the principal planes of the system, f and f' are the focal lengths, x and x' are the conjugate image and object distances, D, s, and L are the dimensions of the vessel, and D', s', and F' are the dimensions of the spectrograph collimator. One should note that the necessary images can be produced by an optical system only if D' and s' have different signs (in the present system D' > 0 and s' < 0). The interior of the vessel is imaged along both sides from the collimator and slit of the spectrograph. The image considered satisfies a basic requirement imposed on the image in Raman scattering investigations [37]. Each light ray which passes simultaneously through the front and back walls of the vessel is used by the spectrograph; this achieves maximum use of the scattered light. The image of the vessel at the spectrograph permits masking of the parasitic light from the side walls, uniform filling of the collimator, and uniform illumination of the slit.

The formulas for the focal lengths and conjugate image distances of an optical system which gives the necessary images have the following form:

$$f = -\frac{F'}{\frac{D'}{D} - \frac{s'}{s}}, \qquad x = f\left(1 - \frac{s}{s'}\right), \quad x' = f\left(1 - \frac{s'}{s}\right). \tag{II.2}$$

The quantities $f = f'$, x and x' are reckoned from the principal planes of the optical system in the direction of the object region or the image region respectively; in Fig. 7 all of them are negative. The quantity s/s' can be calculated from the Lagrange–Helmholtz relation, which is conveniently written in the same form as in [37]:

$$\frac{s}{s'} = -\frac{LD'}{F'D}. \tag{II.3}$$

The ratio s/s' determines the magnification of the slit system and also holds for the slit width: s/s' = h/h'. The formulas given have the same form as in [37]; however, in applying them it is necessary to bear in mind that here, in contrast to [37], D' and s' have different signs.

Depending on the signs of D' and s', the indicated image can be realized either by positive or by negative optical systems. Figure 7 shows the system of images for a negative optical system (f, f' < 0). It is convenient to use a condenser which does not lead to a significant increase of the distance between the vessel and the spectrograph.

A two-lens condenser scheme which gives the image described above is shown in Fig. 8. Here H_1 and H_2 are the principal planes of the first and second lenses of the condenser; d is the distance between lenses; a and b are the conjugate image and object distances; D, s, L, D', s', and F are the dimensions of the mirror system and of its image at the collimator of the spectrograph; and D" and s" are the intermediate images of the back mirror and the slit, produced by the first lens of the condenser. The image s" is a virtual, intermediate, image D" of the back mirror image, located between the lenses of the condenser.

To make the condenser it was necessary to select the lenses to go into it and to calculate the distances a, b, and d. To choose the condenser lenses one must calculate their focal lengths f_1 and f_2 from the following formula:

TABLE 1. Design of the Two-Lens Condenser (Dimensions in mm)

Vessel	Spectrograph	First lens	Second lens	Separation
$L = 200$	$F' = 900$	$f_1 = 116$	$f_2 = 102$	$a = 120$
$D = 20$	$D' = 100$	$\varnothing_1 = 24.8$	$\varnothing_2 = 19.1$	$b = 100$
$s = 8$	$s' = -7.2$			$d = 296$

$$f_1 = \frac{c}{b - x'}, \qquad f_2 = \frac{c}{a - x}, \qquad (\text{II.4})$$

where $c = (a - x - f)(b - x' - f) - f^2$. These formulas derive from formulas (II.5) found below, which are used in the final design of the condenser for the chosen f_1 and f_2. The quantities determined by the general formulas (II.2) of the system are used as calculational data along with the distances a and b of the condenser system, which can be chosen freely. (We chose a and b proceeding from the desired dimensions of the apparatus.) In designing the condenser lens it is also necessary to estimate the distance d between lenses, which, together with a and b, determines the dimensions of the apparatus; the corresponding formulas are given below (II.5).

In practice the focal lengths of the chosen lenses can turn out to be somewhat less than calculated. For the chosen lenses (with focal lengths f_1 and f_2) the final calculation for the condenser is carried out by the following formulas:

$$a = -f + x + f_1 + f_1^2/\Delta, \; b = -f + x' + f_2 + f_2^2/\Delta, \quad d = f_1 + f_2 + \Delta, \qquad (\text{II.5})$$

where a, b, and d are the distances in the condenser,

$$\phi_1 = s + \frac{a}{L}(D + s) \quad \text{and} \quad \phi_2 = s' + \frac{b}{F'}(D' + s') \qquad (\text{II.6})$$

are the diameters of the lenses, and $\Delta = -f_1 f_2/f$ is the optical interval.

In our case the dimensions were determined by the apparatus, and the calculated parameters of the two-lens condenser are given in Table 1.

The condenser lenses were two-component glass achromats. Provided that the relative aperture of the condenser lenses is comparatively small, they provide the high-quality images necessary with the reflections of the mirror system. A diaphragm is placed between the first and second lenses at the position of the image of the back mirror, which excludes light from the walls of the mirror system. The condenser system is shielded from extraneous light by blackened tubes.

§7. Spectral Apparatus and Recorder

In the present work a photographic method was used to record the spectra. A three-prism spectrograph HUET B-III with glass optics was used for recording; the collimator diameter was D' = 100 mm, the focal length of the collimator was F' = 900 mm, and of the camera F_C = 650 mm. The dispersion region of the spectrograph was such that it permitted simultaneous recording of the scattering spectra from both of the excitation lines 4047 and 4358 A. The recording region was sufficient to cover Raman scattering out to 3500 cm^{-1}. To increase the dispersion of the spectrograph, the prisms were changed from minimum deviation; the angle of rotation of the prisms was about 2°. Due to the rotation of the prisms, the dispersion was increased by a factor of about 1.5 and had the following values: at 4047 A, 22.7 cm^{-1}/mm; at 4358 A, 37.1 cm^{-1}/mm and at 4500 A, 42.8 cm^{-1}/mm. The dispersion of the spectrograph changed with temperature; the temperature shift of lines in the 4358 A region amounted to 0.7 cm^{-1}/deg; during photographing the spectrograph was thermostatically controlled with a TC-15 ultrathermostat.

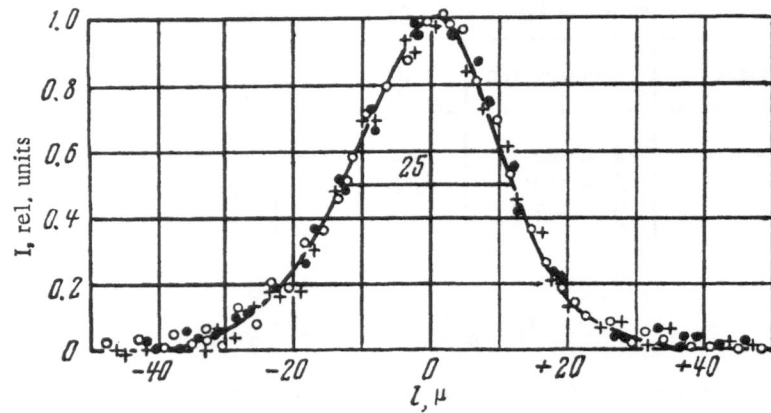

Fig. 9. Instrumental function of coarse-grained photographic plates in
the region from 4000 to 5000 A.

The diffractive limit of resolution for the spectrograph described amounts to 0.3 cm^{-1} at λ = 4358 A, which corresponds to its normal slit width h_N = 5 μ. To estimate the actual resolution of the spectrograph, a measurement of the image breadth of the Hg 4358 A line from a Sosinski lamp was carried out with a microscope. The measured line had a breadth of order 0.3 cm^{-1} and the spectrograph gave an image of it with a breadth of about 0.4 cm^{-1}.

The following kinds of highly sensitive photographic plates were used for recording: Ilford Zenith, Eastman Kodak II-O, Kodak Limited O-aO and Raman Ortho Agfa. The sensitivity of the plates for many-hour exposures was increased by preliminary illumination. The plates were developed in DK-50 developer; the developing time for γ close to γ_{max} was about 10 min at 18°C.

In photographic photometry, the characteristic curves measured on broad lines or continuous spectra are usually employed. The possibility of applying such a method to the measurement of narrow lines is not at all obvious; for narrow lines, in view of the scattering of light in the emulsion and the edge effects, the apparent shape of the characteristic curves can be changed [42].

We carried out an investigation of the relation between shape of the characteristic curves and the breadth of the recorded image. For this purpose the monochromatic 4358 A mercury line was investigated with spectrograph slit widths from 0.005 to 0.5 mm. Analysis of the results showed that for all the types of plate studied, the shape of the characteristic curve does not depend on the breadth of the recorded image. With spectrograph slit widths less than 0.1 mm, a distortion of the image begins, which is connected with scattering of light in the emulsion. Consequently, light which falls on the plate is distributed over an area larger than the optical image. This led to a line shift during measurement of the characteristic curves on such lines. However, measurement of the total blackening of the smeared contour showed that the characteristic curves measured on broad lines (greater than 0.5 mm) remained correct. In the sequel, marks of blackening were recorded from the continuous luminescence spectrum of quinine, excited by a UFO-2 mercury lamp.

Because of the scattering of light in the emulsion layer and the granularity of the plates, a distortion of the shape of the image occurs during recording, which is already noticeably developed on lines which have a breadth less than 0.1 mm. The author, together with A. V. Rakov, investigated the distortion of the recorded line contours on the plates and developed a method for studying it. Quantitatively, the distortion of the investigated line contour was usually considered to be an instrumental function, which describes the blurring of an infinitely sharp line during its recording. To determine the blurring introduced by the plate, the observed line contours of the 4047, 4358, and 4900 A lines of a Sosinski lamp were studied. These lines have a very small true breadth: $2\gamma \approx 0.3$ cm^{-1}. With a normal spectrograph slit, the optical image of the lines on a plate had a breadth around 0.4 cm^{-1} (8 to 10 μ).

The images were recorded on plates of the kind indicated above. The photographs of the lines were measured on an MF-4 microphotometer with a photometer slit of 5 μ. The results of the measurements were converted to intensities by means of the characteristic curves. The observed contours of the investigated lines were plotted in intensity, and will be considered to be an instrumental function of the plates (Fig. 9).

The investigation showed that at the maximum of the image of an infinitely sharp line, over 0.2 to 2.0 breadths the shape of the instrumental function contour remained constant with a change in the photographic density (the various points on Fig. 9 correspond to recording of the instrumental function in the indicated photographic region). The contour of the instrumental function is flattened at the maximum; the wings begin to fall off exponentially at 3 μ from the center; the breadth of the contour is 25 μ. The fundamental contribution to its breadth (15 to 20 μ) is due to the scattering of light in the photographic layer. The shape of the wings agrees with the exponential decay of the light scattering in the emulsion. The flattening of the instrumental function at the maximum is connected with the shape of the recorded images and the granularity of the plates. The small asymmetry in the instrumental function contour is explained by the angle of incidence of the rays which form the image. The investigation of the instrumental function showed that in the wavelength region from 4000 to 5000 A, the breadth and shape of the contour (in mm) does not change. For all the kinds of photographic plates used, the instrumental function remains the same. The kinds studied (highly sensitive coarse-grained materials) have a similar emulsion structure and this explains the results obtained.

Measurement of the observed contour for various spectrograph slit widths showed that with a 20 μ image breadth or less, the observed contour agrees fully with the instrumental functions measured earlier. This shows that the measured instrumental function is determined only by the properties of the photographic layer. In magnitude, the instrumental function significantly exceeds the limit of resolution of the spectrograph and in fact determines the resolution of recording.

The investigated instrumental function was used to study the distortion due to the spectrograph and plates. Exclusive of distortion, its contour had a dispersion $2\gamma_i = 25$ μ. Converting to a scale of inverse centimeters, its breadth for our apparatus had the following values: at 4047 A, $2\gamma_i = 0.8$ cm^{-1}; at 4358 A, 1.0 cm^{-1}; and at 4500 A, 1.3 cm^{-1}.

§8. Photographic Recording of the Spectra

Technical gases were used for the investigation, which were cleaned by silica gel wadding before being admitted to the illuminator. During study of the line contours, the spectral width of the slit was set at $\frac{1}{3}$ to $\frac{1}{4}$ of the breadth of the line investigated. On each plate the Raman scattering spectrum, the excitation lines, and the marks of blackening were photographed. Raman scattering spectra were recorded for both of the excitation lines 4047 and 4358 A. The excitation lines were photographed with the same spectrograph slit width as the spectrum of the gas. Their contours were used to study the distortion introduced into the investigated scattering spectrum during photographing. The exposure for the investigated lines was selected so that their photographic density D lay in the region $0.5 \leq D \leq 1.5$. The exposure for photographing a gas was from 3 to 40 hr. For statistics three or four photographs were made at each pressure point. The random shift of the lines during photographing was controlled by the sharp lines of mercury in the scattering spectrum; if a shift was present the photographs were discarded.

The photographs obtained were measured on an MF-4 microphotometer. The photometer slit was about $\frac{1}{5}$ of the breadth of the lines under investigation; this eliminated the need to study the distortion introduced into the shape of the contour. The background in the spectrum was measured at distances significantly exceeding the breadth of the investigated line, on both sides of it. From the photometer measurement data and the characteristic curve, the line contour (of intensity) was constructed. This operation was carried out on a special transformation stage constructed by the author.

CHAPTER III

INVESTIGATION OF THE ROTATIONAL BANDS
OF OXYGEN AND NITROGEN GAS
IN RAMAN SCATTERING
AT VARIOUS PRESSURES

§1. Investigation of the Rotational Band of Oxygen in Raman Scattering at Various Pressures [43]

The rotational bands of O_2 and N_2 arise from transitions between rotational levels of the basic vibrational states. The transition $\Delta J = 0$ gives the central component, corresponding to the Rayleigh line; the transitions $\Delta J = \pm 2$ give the O and S branches located on both sides of it. For O_2 and N_2 these branches form a spectrum of nearly equidistant lines which extends at room temperature to 150 to 200 cm^{-1} on both sides of the excitation line. Investigation of the rotational bands of O_2 and N_2 in Raman scattering was carried out as a function of the intrinsic pressure of the investigated gas.

The rotational bands of oxygen were investigated in the gas at 27°C over the pressure range from 7 to 125 atm. The 400 atm illuminator with PRK-2 lamps was used to photograph the spectrum. The recording conditions have been described previously (see Chapter 2, §8). Photographs of the observed rotational band in oxygen, with the excitation line Hg 4047 A, are shown in Fig. 10 for pressures of 15 and 75 atm. There are no components corresponding to transitions with even J in the rotational spectrum of oxygen, because of the forbiddenness connected with the nuclear spin value. The separation between the observed components is $\Delta_\omega \approx 11.5$ cm^{-1}. At room temperature a substantial number of rotational levels are populated; in the O branch

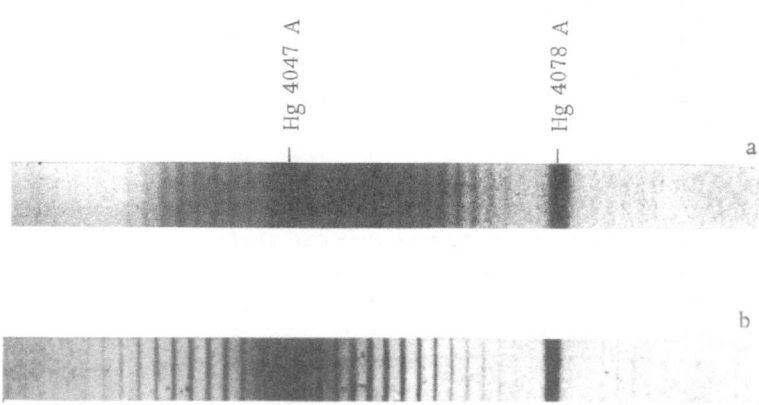

Fig. 10. Rotational band of oxygen in Raman scattering from the 4047 A excitation line at 75 atm (a), and 15 atm (b) pressure.

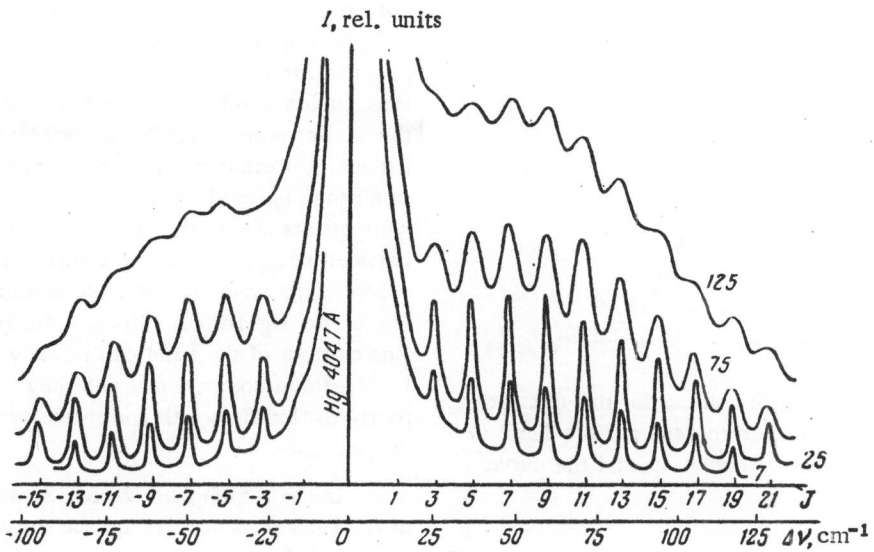

Fig. 11. Rotational band spectra of oxygen in Raman scattering at pressures 7, 25, 75, and 125 atm, at 27°C.

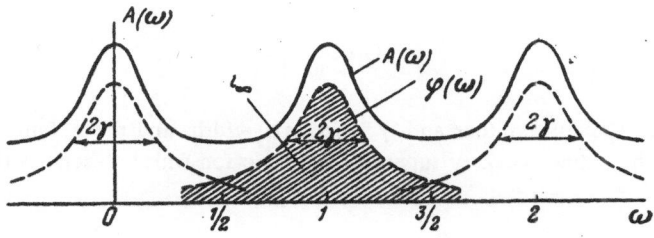

Fig. 12. The Airy function.

of the rotational band, components were observed corresponding to transitions with J = 1 to J = 25, and in the S branch from J = 1 to J = 35.

At low pressure the lines in the rotational band were sharp, and with increasing pressure they were broadened. At 25 atm they began to overlap, and above 75 atm they merged into a continuous band with small maxima (Fig. 11).

Measurement of the line frequencies in the rotational band of oxygen showed that over the whole pressure range studied they did not change within the errors of measurement (± 0.2 cm^{-1}), in agreement with the values calculated from the nonrigid rotator model. The following formula [44] was used to calculate the frequencies:

$$\nu_\omega = \pm \; [(4B_e - 6D) \, (J + {}^3/_2) - 8D \, (J + {}^3/_2)^3], \qquad (\text{III.1})$$

where B_e is the rotational constant, $D = 4B_e^3 / \omega_e^2$ is a constant which describes the nonrigidity of the rotator, and ω_e is the vibrational constant. For the ground state ($^3\Sigma_g^-$) of O_2 these constants have the following values [44]: B_e = 1.4456 cm^{-1}, ω_e = 1580.36 cm^{-1} and D = 4.82 $\cdot$ 10^{-6} cm^{-1}.

To determine the breadth and intensity of the individual components of the rotational spectrum it was necessary first to separate them from the observed band, and then to separately observe the contours to determine the true values of the breadth and intensity.

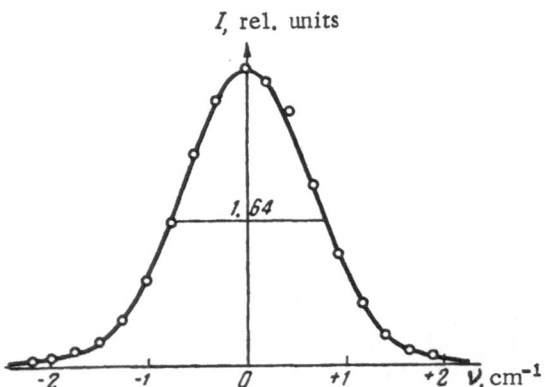

I, rel. units

1.64

-2   -1   0   +1   +2 ν, cm^{-1}

Fig. 13. Overall distortion contour for the excitation lines Hg 4047 A with a spectrograph slit h = 1.25 cm^{-1}. The points are experimental, and the curve is Gaussian.

Separation of the components in the band amounts in essence to eliminating the effect of the neighboring components. In the separation, each component of the band is considered as an independent line; such a point of view is justified provided that the components of the band correspond to transitions between different levels of the molecule. In the pressure range up to 20 atm, the breadth of the separate components is significantly less than the separation between neighboring components of the band; the observed distortion of the band contour due to merging is not significant. In this case the observed line contour of the band can already be considered to be separated components, and they can be used directly to find the width and intensity of the investigated lines.

In the region from 20 to 125 atm, the breadth of the lines does not exceed the separation between them and the individual lines are distinctly revealed in the observed band. A method based on use of the Airy function [45] was employed to separate the individual components in this region. The Airy function (Fig. 12) is the result of superposing equidistantly spaced components of equal intensity, which have a Lorentzian shape. It is defined in the following way: if the component functions have the form

$$\varphi(\omega) = \frac{i_\infty \gamma/\pi}{\omega^2 + \gamma^2},$$ (III.2)

where i_∞ is the area of a component contour and γ is its half-width, and these functions are spaced at a distance $\Delta\omega = 1$ from each other; then the Airy superposition function which describes them is defined by the expression

$$A(\omega) = \sum_m \frac{i_\infty \gamma/\pi}{(\omega - m)^2 + \gamma^2} = \frac{i_\infty \coth \pi \gamma}{1 + \dfrac{\sin^2 \pi \omega}{\sinh^2 \pi \gamma}}.$$ (III.3)

Analysis of the rotational band of oxygen at 40 and 60 atm showed that separate sections of it, which include several lines, are well approximated by the Airy function, and that components separated by the method of successive approximations have the Lorentzian shape. This permits use of the Airy function to determine the observed breadth and intensity of the individual components. The error in the determination of the observed breadth of the rotational lines amounts to about 20%, and for the integrated intensity is about 5%.

The components separated from the band have the shape observed in isolated lines in Raman scattering. This shape includes the true breadth of the investigated line and the instrumental distortions, i.e., the finite breadth of the excitation line, the width of the spectrograph slit, and the instrumental function of the spectrograph and the detector. The theory of reduction [46] considers how to take into account the distortion during recording and how to find the true shape of the investigated line. According to this theory, each distortion can be described by an instrumental function a(x). During recording, each element of the real distribution I(x) is smeared into a blur determined by the instrumental function, due to the distortion. In this way the observed distribution $f(x)$ is obtained. The relation between the observed contour $f(x)$ and the real distribution I(x) is described by a convolution integral [46]:

$$f(x) = \int_{-\infty}^{\infty} I(x') a(x - x') \, dx'.$$ (III.4)

168

If there are several distorting factors they can be successively superimposed on each other by means of this convolution. The observed contour is produced by the superposition of all the distortions. The properties of the convolution provide an arbitrary way of interchanging and grouping the distortions. This permits simultaneous consideration of several distortions described by their overall instrumental function, and allows them to be included in an arbitrary order.

The reduction theory was used to analyze our results. In order to study the distortion, the excitation lines were photographed with the same spectrograph slit as the Raman spectrum during the study of the rotational spectrum of oxygen. The contour λ^* measured during this recording included all of the distortions. The rotational spectra were located close to the excitation line, and changes in the spectrograph dispersion over this region were small and were not considered during the analysis. To find the true Raman scattering line breadth, the overall distortion contour λ^* was directly eliminated from the observed contour γ^* of the investigated line. The method of elimination, defined by the convolution equation, depends on the shape of the contour under consideration. In our case the Raman scattering lines were broad and had a Lorentzian shape. The overall distortion in λ^* had contributions from the excitation line, the spectrograph slit, and the instrumental function of the apparatus. The observed shape of the λ^* contour (intermediate between Lorentzian and square) was approximately Gaussian. This approximation is supported by the experimental measurements of the λ^* contour, shown in Fig. 13. The observed breadths of the contour of the overall distortion for PRK-2 lamps and a 1.25 cm^{-1} spectrograph slit was 1.64 cm^{-1} for the 4047 A region and 1.89 cm^{-1} for the 4358 A region. The mean deviation of the measurement of the breadth of the instrumental distortion contour was 10%.

The observed contour was the convolution of Lorentzian and Gaussian curves, and is described by the Voigt function. The relation between the widths of the curves for such a superposition has been analyzed by S. G. Rautian [46] and by G. G. Petrash (present collection).

The experimental investigation showed that the line breadth in the rotational spectrum of oxygen depends linearly on the pressure over the whole region studied, from 17 to 125 atm (Fig. 14); the relative breadth is $\gamma_\omega / p = 0.055 \pm 0.005$ cm^{-1}/atm (γ_ω is the breadth of the rotational lines). For all pressures studied, γ_ω did not depend on J ($3 < J < 15$) within the experimental error (15 to 20%), i.e., at a given pressure the broadening was the same for all of the lines in the rotational spectrum (Fig. 15).

If the results of more recent work [47] are used, where the breadth of the rotational lines of oxygen was measured in the region from 3-10 atm, the $\gamma_\omega(p)$ curve of Fig. 14 can be extrapolated to $p = 0$. In the given case, the absence of residual breadth of the rotational lines at $p = 0$ showed that the basic (and sole) cause of the observed broadening of the rotational lines is molecular collision broadening. In addition to the collision broadening, for Raman scattering lines in a gas, intrinsic and doppler broadening were observed. Estimates relating to this broadening have been made by Kh. E. Stepin [33]. They show that the broadening introduced by these processes does not exceed 0.03 cm^{-1}, and thus lies within the limits of error of the experiment.

The Lorentzian line shape observed in the experiment, the linear dependence of $\gamma_\omega(p)$, and the absence of overlap of the rotational levels permits application of the theory of collision broadening of spectral lines [1, 2] to the rotational band. According to the collision theory, the line breadth 2γ is given by relation (I.1) or, if the relation is expressed in reciprocal centimeters,

$$2\gamma \, [\mathrm{cm}^{-1}] = \frac{N v_{rel} \rho^2}{c} , \qquad\qquad (\mathrm{III.5})$$

where $v_{rel} = (16RT/\pi M)^{1/2}$ is the relative velocity of the colliding molecules, c is the velocity of light, and ρ is the Weisskopf radius. For oxygen at 300°K ($v_{rel} = 6.3 \cdot 10^4$ cm/sec) this relation leads to the following formula for the relative broadening

$$\frac{2\gamma}{p} = \frac{L v_{rel}}{c} \rho^2 = 5.64 \cdot 10^{-3} \rho^2, \qquad\qquad (\mathrm{III.6})$$

TABLE 2. Experimental Data on Broadening of the Rotational Lines in O_2

Region of measurement	Pressure range	γ_ω/p, cm^{-1}/atm	ρ_ω, A	Reference	
Raman scattering	7—125 atm	0.055	4.43	Author	[43]
	1—10 atm	0.0504	4.35	Bazhulin and Lazarev	[47]
Microwave absorption	1—8 torr	0.0505	4.35	Anderson	[24]

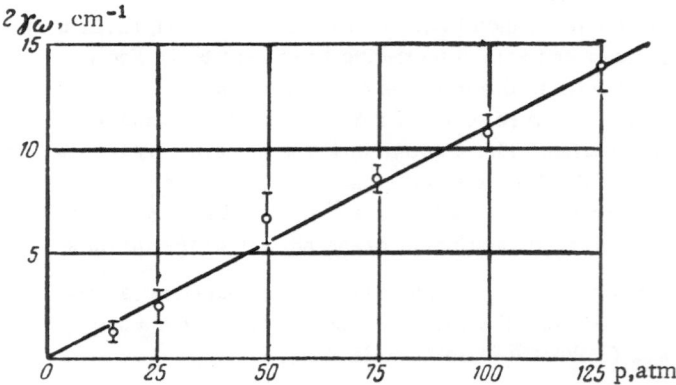

Fig. 14. Line breadth versus pressure for the rotational band of oxygen in Raman scattering.

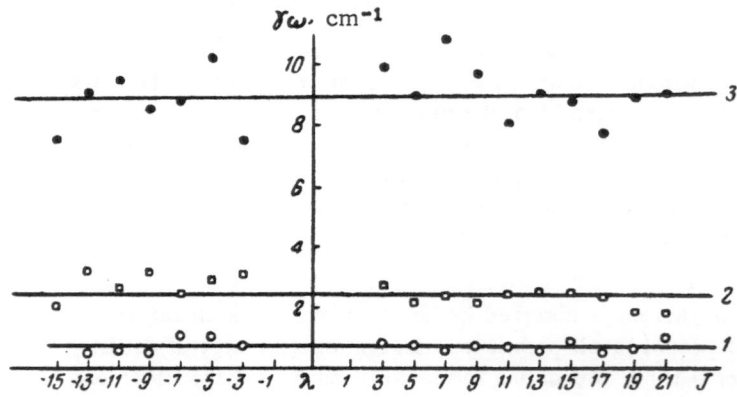

Fig. 15. Line breadth versus J, for the rotational band of oxygen at various pressures. 1) 7 atm; 2) 25 atm; 3) 75 atm.

where 2γ is the full line breadth (in cm^{-1}), L is the Loschmidt number, p is the pressure (atm), and ρ is the Weisskopf radius (in A). Using the experimental value $\gamma_\omega/p = 0.055$ cm^{-1}/atm taken from Fig. 14, we obtain a value of 4.43 A for the effective radius of the collisions which produce the broadening of the rotational lines of O_2. The measured value of ρ_ω is somewhat larger than the gas-kinetic value of the collision radius of O_2, for identical molecules of equal gas-kinetic diameter $d_0 = 3.61$ A [1]. The difference is reasonable, since these quantities characterize different processes.

The measured value of ρ_ω allows one to determine an upper limit for the region where collision broadening acts, determined by the collision theory condition (L2): $2\pi\rho^3 N \ll 1$. Hence it follows that for the rota-

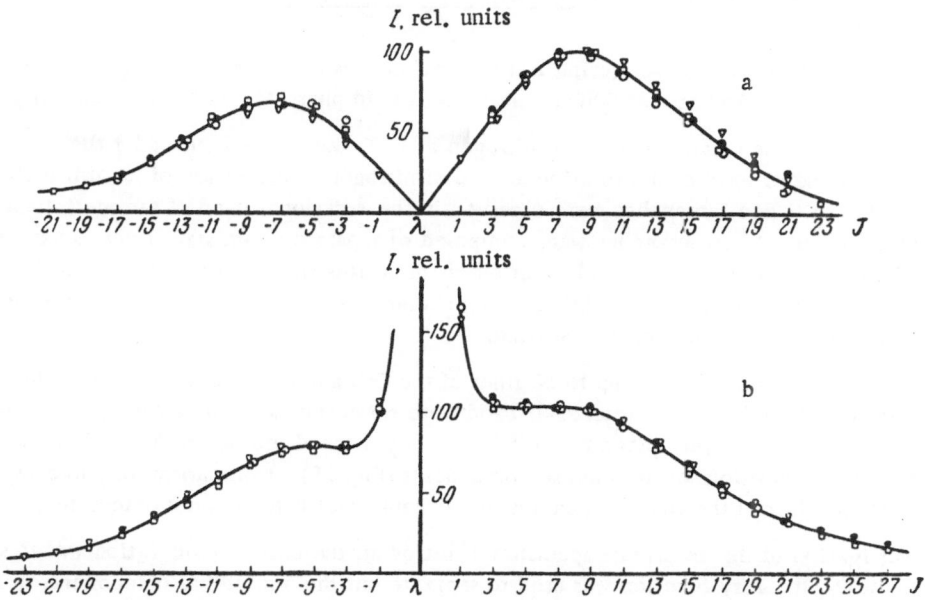

Fig. 16. Distribution of intensity over the rotational band of O_2. a) At pressures 7, 15, 25, and 50 atm with the excitation line excluded; b) at pressures 50, 75, 100, and 125 atm with the excitation line included. The different points refer to measurements at different pressures.

tional levels of the ground state of O_2, the collision theory is valid in the region p < 70 atm, i.e., the pressure investigated lie within the region of applicability of the collision theory, and the maximum pressure lies on the edge of it. Unfortunately, because of the complete overlap of the rotational lines it was not possible to investigate the breadth at higher pressures.

Our results agree well with the literature data for broadening of the rotational lines of O_2 in Raman scattering at pressures from 1-10 atm [47], and absorption in the microwave range at pressures 1-8 torr [24]. Both of these investigations were carried out at room temperature. These results are compared with our data in Table 2 (in which the average value of γ_ω for lines in the maximum intensity of the band is given). The agreement of the relative broadening γ_ω/p and the dependence of the breadth on J for O_2 in microwave absorption and Raman scattering allows us to assume that in both cases the interaction mechanism which determines the broadening is the same.

The distribution of intensity over the individual lines in the rotational band is determined by the thermal occupation of the levels and their statistical weights [44]

$$\mathcal{I} = C \left(J + J' + 1 \right) e^{-BJ \, (J+1) \frac{hc}{kT}} , \tag{IIL 7}$$

where J and J' are the initial and final states and C is a constant of the molecule. Investigation of the distribution of integrated intensity over the individual lines of the rotational band shows that this distribution is actually given by the formula and does not change over the whole investigated pressure range (Fig. 16).

A redistribution of the intensity in the rotational band can be observed due to the transfer of intensity into the central line [3]. However, this can take place only after overlapping of the levels. It can be shown that for the pressure range investigated, the greatest observed broadening remains less than the distance between rotational levels and there is no level overlap. This is explained by the preservation of the intensity distribution in the band with a change of pressure.

§2. Investigation of the Rotational Band of Nitrogen in Raman Scattering at Various Pressures [48]

In the present work, the Raman scattering in the rotational band of gaseous nitrogen was investigated from 7 to 114 atm at 27°C, at 150 atm at 200°C, and in the liquid phase at −196°C and normal pressure.

The investigation of the rotational band of nitrogen at 27°C was carried out under the same conditions as with oxygen. The general form of the rotational band of nitrogen at a pressure of 27 atm is shown in Fig. 17. As with oxygen, only the O and S branches were observed in the rotational band of nitrogen, located on both sides of the excitation line. These branches were composed of separate equidistant lines, separated by $\Delta \nu_\omega$ = 8.0 cm^{-1}. In the branches of the band an alternation of intensities was observed, determined by the nuclear-spin value; the intensity ratio of neighboring components was 1 : 2. There was the usual thermal distribution of intensity over the components of the band.

In the spectrum of the gas at 27°C, up to 15 lines of the O branch and up to 25 in the S branch were observed. At low pressure the lines in the rotational band were sharp, but with increasing pressure they broadened. At 20 atm they began to overlap and above 60 atm they merged into a continuous band with weakly developed maxima at the positions of the intense components (Fig. 18). Comparison of photographs at different pressures shows that any shift of the line frequencies in the rotational band was not detectable.

During investigation of the rotational spectrum of nitrogen, the basic consideration was to study the line breadth. As for O_2, the Airy function was used to separate out the individual components in the rotational band of nitrogen. The distortion was taken into account and the true parameters of the investigated lines found on the basis of the reduction theory, considered in detail in Section 1 of this chapter.

The experimental pressure dependence of the breadth of the rotational lines of N_2 is given below (the average breadth for transitions with J = 10 to 15 is given, i.e., in the maximum intensity part of the band).

p, atm	$2\gamma_\omega$, cm^{-1}	p, atm	$2\gamma_\omega$, cm^{-1}
7	1.75 ± 0.9	40	5.8 ± 0.8
15	2.2 ± 0.7	60	7.1 ± 1.2
25	4.7 ± 0.7	80	7.5 ± 1.5

A linear pressure dependence is observed for the line breadth in the rotational band of nitrogen at 27°C in the region up to 50 atm; the relative broadening is γ_ω/p = 0.073 cm^{-1}/atm. In this pressure region the density of molecules is comparatively small and the collision broadening theory can be applied. The effective collision radius for lines in the rotational band of N_2, determined from the relative broadening, is 4.9±1 A. This value agrees within the limits of experimental error with the value ρ_ω = 3.95 A, obtained in [47] for the Raman scattering lines in the rotational band of N_2 for pressures from 3-10 atm. As for oxygen, the effective broadening radius for Raman scattering lines in the rotational band of N_2 is somewhat larger than the gas-kinetic diameter of N_2, 3.75 A [1].

The upper pressure limit for the region of collision broadening of the rotational lines of N_2 determined from ρ_ω amounts to 50 atm. Thus for N_2 the collision broadening involves a linear relation between breadth and pressure.

In the higher pressure region (above 50 atm) the experimental points lie below the curve which expresses the relation $\gamma_\omega(p)$. The lines of the rotational spectrum strongly overlap here and application of the Airy method does not take into account the falling off the intensity in the wings of the band, which leads to results which are too low. It is also difficult to apply the general calculation of the band shape here, because it is not clear how to take into account the background from the excitation line. In addition, it could be that statistical broadening appears here; this is indicated by an estimate of the limit of collision broadening based on the value of ρ_ω found. This could also lead to a decrease of the experimentally determined values of the breadth. All of these circumstances could cause the observed decrease of the relative broadening with p > 50 atm.

Fig. 17. Photograph of the rotational bands of nitrogen in Raman scattering, observed with the Hg 4047 A excitation line at 20 atm pressure.

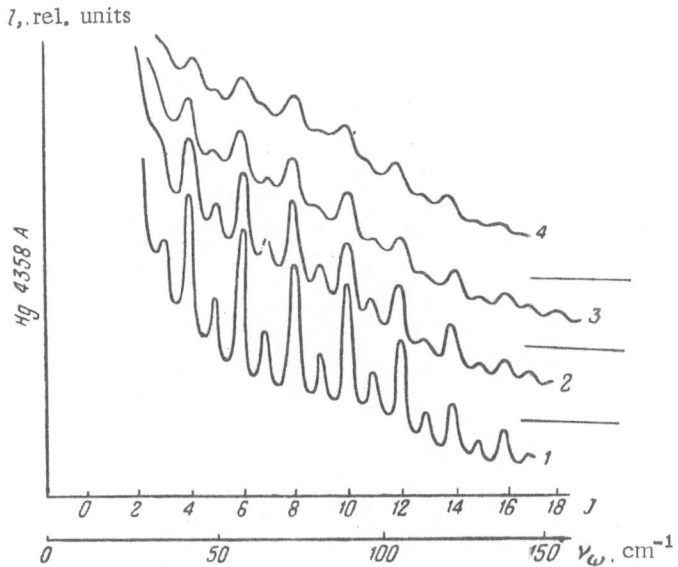

Fig. 18. Intensity distribution in the S branch of the rotational band of nitrogen at 27°C. 1) 15 atm; 2) 25 atm; 3) 40 atm; 4) 60 atm; the straight line gives the zero of intensity for the corresponding pressure.

In studying the collision broadening, there is considerable interest in the changes of the optical collision cross section σ with temperature, since the nature of the interaction during individual collisions is changed. An investigation was carried out in the gas at 200°C, in the apparatus described in Section 4 of the following chapter. With 150 atm at 200°C, up to 15 lines of the Q branch and 25 in the S branch were observed in the rotational band of nitrogen. Already there was an observable change in the distribution of intensity in the band; the rotational band was broader than at 27°C. The broadening of the band relative to the band at room temperature is connected with a change in the thermal population of the rotational levels. The lines in the rotational band strongly overlap and have a breadth $2\gamma_\omega \approx 9.0 \pm 2$ cm^{-1}, i.e., of the same order as for 90 atm at 27°C ($2\gamma_\omega \approx 8.5$ cm^{-1}). This is the pressure at which the density of molecules is the same. No significant dependence of the breadth on temperature was observed, and in general this does not contradict the collision broadening theory. Actually, the collision frequency decreases with temperature and also the interaction at an individual collision, which determines the optical collision cross section σ. These factors can mutually compensate on the whole, so that a change in broadening with temperature might not be observed.

The spectrum of liquid nitrogen was investigated at $-196°C$ and normal pressure. The conditions for photographing the spectrum are described in Section 4 of the following chapter. In the case of liquid nitrogen it is of interest to be able to detect the rotational band.

However, as in [49], it was not discovered. In view of the substantial drop in temperature, the rotational band shrank into the Rayleigh line and its wings appeared to be strongly masked. Because of this it was not observed in the liquid.

CHAPTER IV

INVESTIGATION OF THE Q BRANCH
OF THE VIBRATIONAL BAND
OF O_2 AND N_2 IN RAMAN SCATTERING

§1. Structure of the Q Branch of the Vibrational Band of O_2 and N_2 in Raman Scattering

The nature of the sizable observed breadth of the vibrational lines in Raman scattering remained an open question up to the present work. In the discussions in review articles [33, 34] (see Chapter 1, §2B) it is indicated that the reason for the breadth can be both unresolved structure lines and broadening due to intermolecular interaction. Investigation of both these causes is the basic direction of the present study of the vibration line contours in Raman scattering.

The vibrational bands of O_2 and N_2 in Raman scattering arise from the transitions $\Delta v = +1$, $\Delta J = 0, \pm 2$. For $\Delta J = 0$ transitions form the central Q branch of the band, and the $\Delta J = \pm 2$ transitions produce the O and S branches located on both sides of the Q branch. In structure, the O and S branches precisely duplicate the rotational band $\Delta v = 0$, $\Delta J = \pm 2$, considered in Chapter 3.

The Q branches of O_2 and N_2 possess a structure which arises from the change of the moment of inertia of the molecule in the presence of vibration. The frequencies of the structural components of the Q branch of the vibrational bands of O_2 and N_2 are given by the formula [3]

$$v = v_0 - \alpha_e \, J \, (J + 1), \tag{IV.1}$$

where $\alpha_e = B_e - B_1$ is the vibration−rotation interaction constant.

As in the case of the rotational spectrum, the effect observed here is related to the value of the nuclear spin. For O_2 there are no components corresponding to transitions with even J, and for N_2 an alternation of intensity is observed, in which the intensity ratio of neighboring components is 1:2. The structure of the Q branch of O_2 at 300°K is shown in Fig. 21, where the individual lines correspond to components which arise from transitions with different rotational levels.

The distribution of intensity over the components, as in the rotational bands, is determined by the statistical weights and thermal populations of the levels. As can be seen from formula (III.7), the shape and size of the band depend strongly on temperature.

The separation of the components in the Q bands of O_2 and N_2 is not large (of the order 0.1 cm^{-1}). In recording, the components of the structure are usually not resolved and they merge, forming a continuous band of asymmetric shape. The half-width of the whole structure of the Q branch of O_2 at 300°K is about 2.5 cm^{-1}. Such structure, unresolved by the apparatus, could be the reason for the finite breadth of the observed Q branch.

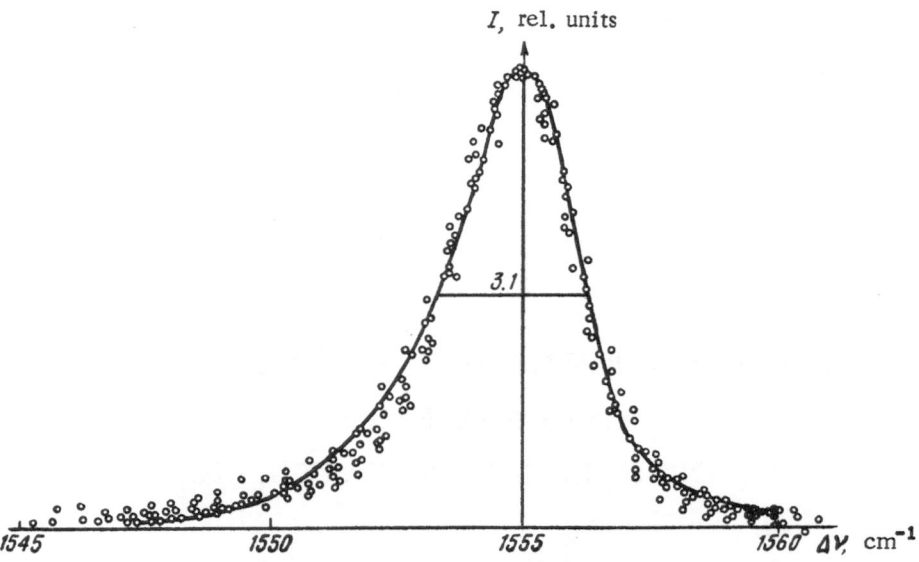

Fig. 19. Observed contour of the Q branch of the vibrational band of oxygen from the 4358 A excitation line, at pressures from 15 to 125 atm. The continuous curve is the calculated contour and the points were measured at various pressures.

§2. Investigation of the Q Branch of the Vibrational Band of Oxygen as a Function of Pressure

The vibrational band of oxygen in Raman scattering was investigated at 27°C in the pressure range from 15 to 125 atm. For this investigation, the same apparatus was used as for taking the rotational band. Since the vibrational lines of O_2 are sharp, in photographing the spectrograph slit was approximately two-thirds of the observed breadth of the excitation line (h = 1.25 cm^{-1}). The exposure time, depending on the pressure, was from 3 to 40 hr. To study the instrumental distortion, the Raman scattering spectrum was recorded simultaneously along with the excitation line, with the same spectrograph slit.

In the vibrational band of oxygen, one sharp line with a frequency ν = 1555 cm^{-1} was observed, corresponding to the Q branch. The low-intensity O and S branches are usually not observed; with substantial exposure and pressure there is a weak blurred wing at the position of the S branch.

Measurements show that the Q branch in oxygen takes the form of a sharp line of asymmetric shape with a wing on the violet side (Fig. 19). The observed breadth (around 3 cm^{-1}) and shape of the Q branch contour do not change over the whole pressure range investigated. The relation between breadth of the Q branch contour and pressure is shown in Fig. 20. No pressure shift of the frequency of the Q branch was observed.

To analyze the Q branch, a calculation of its contour was carried out under the assumption that the Q branch possesses the structure considered in the previous paragraph, and that the individual components have finite breadth. They represent independent lines which correspond to transitions between different levels. If the components possess a finite breadth $2\gamma_c$ then the total contour is obtained by simply superposing them on each other (Fig. 21). The values of frequency and intensity for the components were calculated by formulas (IV.1) and (III.7). The following values were used for the constants of O_2 [44]: α_e = 0.0158 cm^{-1}, B_e = 1.4456 cm^{-1}, ν_0 = 1556 cm^{-1}, and T = 300°K.

The contours of the individual components include both their intrinsic breadth and the instrumental distortion. Since there is no collision broadening of the Q branch, it can be assumed that the contours of the individual components are basically determined by the instrumental distortion. With a finite spectrograph slit width, the instrumental distortion is most closely approximated by a Gaussian curve, for both excitation lines

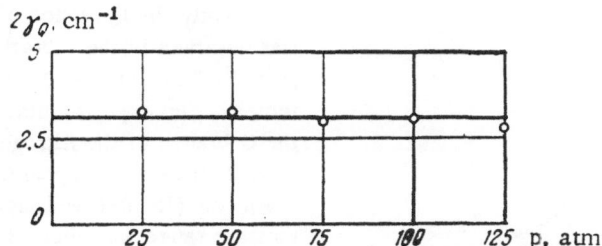

Fig. 20. Pressure dependence of the breadth of the Q branch contour of oxygen, observed from the excitation line Hg 4358 A.

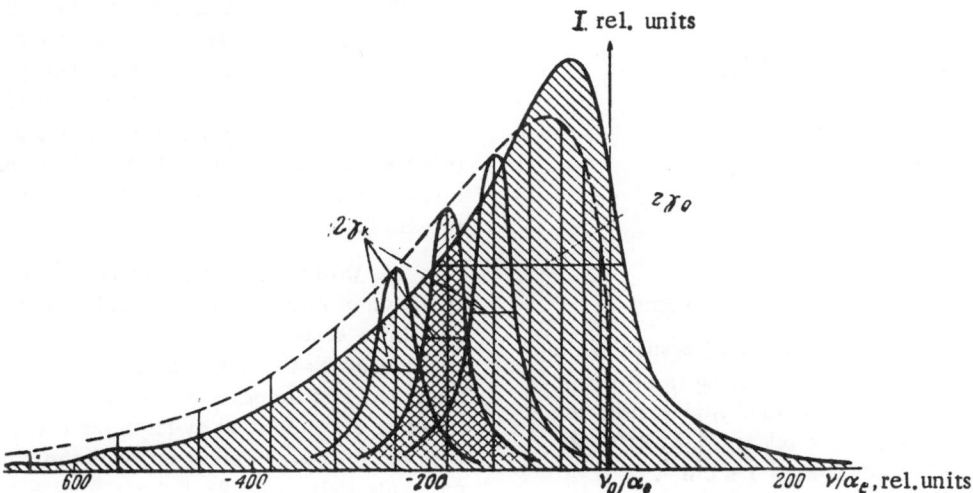

Fig. 21. Structure and calculated shape of the observed Q branch contour of oxygen with finite breadth components.

(see Fig. 13). Calculations of the Q branch contour were carried out for components of Gaussian shape and various breadths. It was assumed that all the components in the structure were broadened identically.

Comparison of the calculated contours with the experimental ones showed that the observed Q branch contour of O_2 (from both the 4047 A and 4358 A excitation lines) is well approximated by a calculation in which the breadth of the components corresponds to the breadth of the instrumental distortion. A comparison of the contours is shown in Fig. 19. The analysis presented for the Q branch of O_2 allows us to assume that the observed contour arises from the structure we have considered, and the finite breadth of the components contained in it. The breadth of the individual components agrees, within the experimental error (± 0.5 cm^{-1}), with the value of the instrumental distortion. Consequently, the intrinsic breadth of the Q branch components of O_2 lies beyond the limit of error of the measurement. The results obtained also confirm the experimentally observed preservation of the Q branch breadth with changing pressure.

The absence of intrinsic breadth of the components and the preservation of the Q branch breadth over the whole investigated pressure range shows that, in contrast to the rotational band $\Delta v = 0$, for the Q branch components of the vibrational transition, collision broadening is practically absent.

§3. Investigation of the Q Branch of the Vibrational Band of Nitrogen at Various Pressures

The effect of pressure on the vibrational band of nitrogen in Raman scattering was investigated at 27°C over the pressure range from 14 to 114 atm. The apparatus and experimental conditions were the same as in the investigation of the vibrational band of O_2.

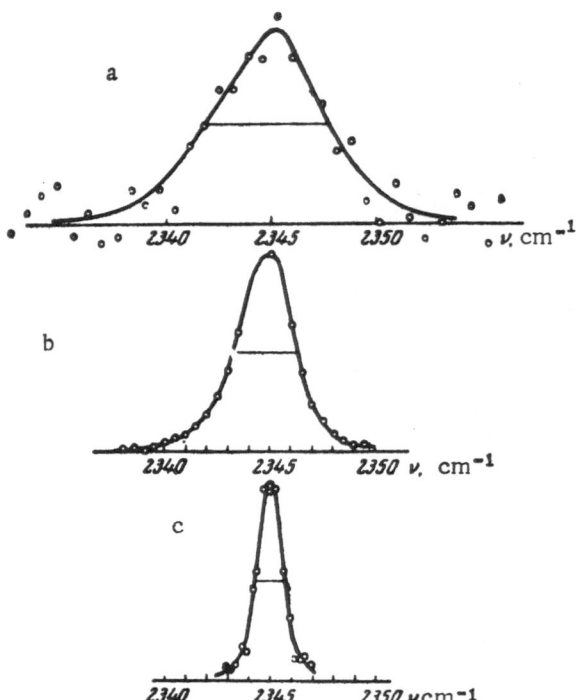

Fig. 22. Observed contours of the Q branch of the vibrational band in Raman scattering in nitrogen from the excitation line Hg 4358 A at various temperatures. a) 200°C, $2\gamma = 5.7$ cm^{-1}; b) 27°C, $2\gamma = 3.1$ cm^{-1}; c) (−196°C), $2\gamma = 1.4$ cm^{-1}.

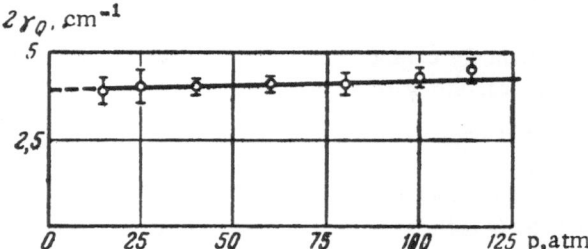

Fig. 23. Pressure dependence of the observed breadth of the Q branch of the vibrational band of nitrogen.

Only the Q branch of the vibrational transition was observed in the vibrational band of nitrogen ($\nu = 2345$ cm^{-1}); the O and S branches were not detected. The experimental measurements show that the Q branch of the nitrogen vibrational band at 27°C (Fig. 22b) has the appearance of a comparatively sharp line (its observed breadth in the investigated pressure range was about 3 cm^{-1}). The line had an asymmetric shape with a wing on the violet side. Its shape did not change over the whole investigated pressure range and agreed with calculation. The dependence of the observed breadth of the Q branch of N_2 on pressure is shown in Fig. 23, from which it is seen that with increasing pressure the Q branch is slightly broadened. However, this change does not exceed 0.5 cm^{-1} over the whole investigated pressure range, i.e., lies at the limit of precision of the measurements. Comparison of the photographs shows that there is no frequency shift of the Q branch in the investigated pressure range with a precision of one or two cm^{-1}.

As with oxygen, the observed contour of the Q branch of nitrogen is explained by the structure and finite breadth of the components. The structure is produced by the same causes as in oxygen. The frequency and intensity of the components of the structure are described by formulas (IV.1) and (III.7) with the following values for the constants of the N_2 molecule [50]: $\nu_0 = 2345.2$ cm^{-1}, $B_e = 1.998$ cm^{-1}, and $\alpha_e = 0.171$ cm^{-1}.

In contrast to oxygen, there is an alternation of intensities for the structural components. This is connected with the different statistical weights of the even and odd levels and is determined by the nuclear spin value of nitrogen (see Chapter 3, §2). Calculations of the observed contour of the Q branch in N_2 were carried out with the same assumptions as for oxygen. In the calculation, each component of the structure was broadened by a Gaussian function with a breadth of 1.6 cm^{-1}. In Fig. 22b the observed contour of the Q branch of nitrogen from the 4358 A excitation line at 60 atm is compared with calculation (the continuous curve is calculated; the points are experimental). The agreement confirms that for nitrogen also, the observed contour is explained by the structure of the Q branch and the broadening of its components.

The breadth of the structural components ($2\gamma_c = 1.6$ cm^{-1}) is determined basically by the instrumental distortions ($2\gamma_i = 1.5$ cm^{-1}). The intrinsic breadth of the components due to collision broadening is not large; even at 100 atm it does not exceed 0.5 cm^{-1}, i.e., lies at the limit of precision of the measurement.

This result agrees well with the results of the O_2 investigation, where no collision broadening of the Q branch (within the limit of experimental error) was observed. The results of an investigation of the Q branch

TABLE 3. Calculated and Observed Breadth of the Q Branch of the Vibrational Branch of N_2 at Various Temperatures

T, °C	State of the substance	Breadth of the structure components $2\gamma_i$, cm^{-1}	Calculated breadth of the contour $2\gamma_Q^*$, cm^{-1}	Observed breadth of the contour $2\gamma_Q$, cm^{-1}
200	Gas, 150 atm	2.5	4.9	5.7 ± 2.0
27	Gas, 90 atm	1.5	3.1	3.0 ± 0.5
−196	Liquid under atmosphere pressure	1.0	1.45	1.4 ± 0.2

of N_2 in the 1-10 atm region [47] (where no collision broadening of the Q branch components was detected) are found to be in full agreement with the present work.

§4. Investigation of the Q Branch of the Vibrational Band of Nitrogen at Various Temperatures [48]

For the Q branch of N_2, in addition to the investigation at 27°C, measurements were made at 200°C and −196°C.

At 200°C the Q branch was investigated at p = 150 atm under the same conditions as for the rotational band at 200°C. In this case an apparatus with internally mounted PRK-2 illuminating lamps was used. The operating regime of the lamps was close to normal and the excitation lines were broader than in the illuminators for investigations at room temperature. The exposure time was 125 hr. With such an exposure, significant errors of measurement are inevitable. The breadth of the overall instrumental distortion function and the precision of measurement during recording was $2\gamma_i = 2.5 \pm 1.5$ cm^{-1}. The observed contour of the Q branch of N_2 at 200°C and 150 atm (see Fig. 22a) repeated the shape of the contour at 27°C (an asymmetric contour with a shadow on the violet side) but its breadth was substantially larger: $2\gamma_Q = 5.7 \pm 2$ cm^{-1}. No frequency shift of the Q branch was observed.

The Q branch was investigated in liquid nitrogen at the boiling temperature (−196°C). To photograph the Raman scattering, a two-lamp illuminator with cylindrical lenses was used, which was developed by the author together with A. V. Rakov [51]. The spectral breadth of the spectrograph slit was 0.75 cm^{-1}. Using a Sosinski lamp, the instrumental distortion for such a slit is determined by the instrumental functions of the spectrograph and plate, as described in Chapter 2. The breadth of the instrumental function in the region of the investigated Raman lines is $2\gamma_i = 1.0$ cm^{-1}.

The observed contour of the Q branch of the vibrational band in liquid nitrogen (see Fig. 22, curve c) exhibits a sharp line. Its breadth is $2\gamma_Q = 1.4 \pm 0.2$ cm^{-1}, i.e., somewhat larger than the value $2\gamma_i$ of the instrumental function.

The results of the investigation of the Q branch of the vibrational band of nitrogen at different temperatures shows that the breadth $2\gamma_Q$ of the contour changes sharply with a change in temperature; the shape of the Q branch contour (asymmetric and shadowed) remains the same over the whole investigated temperature range. There is no shift of the Q branch frequency (with a precision of ±2 or 3 cm^{-1}). The observed change of the Q branch breadth is closely connected with the thermal distribution of intensity in its structure. The breadth of the band which forms the structure is determined by the intensities of its components (see III. 7) and depends on the distribution of the molecules over the rotational levels. With increasing temperature the population of all the higher levels increases, and all the higher components in the Q branch appear; the breadth of its structure increases. A change of the structure with temperature could be one of the basic reasons for the experimentally observed changes in the Q branch. We have calculated the observed contour of the Q branch of the temperatures investigated. The calculation was done in analogy with those done earlier (see §2 of this chapter); we proceeded from the assumption that the broadening of the structural components arises only due to instrumental distortion: $2\gamma_c = 2\gamma_i$.

The results of the calculation are compared with experiment in Table 3, where it is evident that the observed breadth agrees with that calculated, within the experimental error. This indicates that over the whole region investigated, the breadth of the Q branch is fully explained by the change of structure with temperature and the broadening of its components due to instrumental distortion. An analysis of Table 3 also shows that the difference between the breadths of the calculated and experimental contours, which is determined by the intrinsic breadth of the components, is significantly less than the experimental error. Thus the intrinsic breadth of the components of the Q branch practically vanishes.

One should note that for $-196°C$ the experimental results refer to the liquid; in this case the presence of rotationally chaotic motion in the liquid is sufficient to preserve the rotational splitting [34]. This allows the observed breadth of the Q branch to be explained by rotational splitting.

§5. Discussion of the Results

The experimental results show that the regularities of the broadening of the Q branch are identical in O_2 and N_2. The basic cause of the observed breadth of the Q branch in the investigated temperature and pressure ranges is its rotational structure. This determines both the observed breadth and the temperature dependence of the Q branch contour. The collision broadening in this pressure and temperature region practically vanishes for the Q branches of the vibrational bands of O_2 and N_2.

In this respect the Q branch differs fundamentally from the O and S branches investigated earlier in the rotational band ($\Delta v = 0$, $\Delta J = \pm 2$), where a substantial broadening is observed, which is related to the intermolecular interactions.

At the investigated temperatures, the energy of the colliding molecules is still insufficient to perturb the vibrational states; it follows to expect the greatest broadening to be connected with a change of rotation. Actually, the results given in Chapter III do indicate a significant broadening of the rotational levels due to the interaction. The absence of collision broadening in the Q branch while the transitions composing it occur between broadened rotational levels is unexpected. The Q branches of O_2 and N_2 arise from transitions without a change of the rotational state or reorientation of the molecule: $\Delta J = 0$, $\Delta m = 0$. The absence of collision broadening is also indicated by the fact that these transitions do not depend on external perturbations of the rotational states. All this indicates that the vibrational transitions composing the Q branch are completely independent of the rotational state. The transition is pictured as if it is performed by a definite part of a broadened rotational level of the lower state, in precisely the same place as in a level of upper state. The results obtained are new. They indicate yet another property of the vibrational transition which does not have a theoretical interpretation at present.

The investigations of the Q branch of the vibrational bands of O_2 and N_2 allows a more valid approach to understanding the experiment of Kh. E. Sterin, described in Section 2 of Chapter I. The $\nu = 992$ cm^{-1} line of benzene studied by him belongs to the same group of fully symmetric vibrations as the vibrational bands of O_2 and N_2. The absence of collision broadening for lines of the given type immediately allows us to consider that the observed breadth is basically brought about only by the structure of Q branches: this broadening, connected with the structure, is preserved in the liquid, also. Since the change of temperature in the experiment was small, the breadth of the structure remained unchanged. This also explains the preservation of the breadth of the investigated lines during the phase transition, in stephins experiments.

The small sensitivity to intermolecular interactions of the breadth of the Q branch of fully symmetric vibrations is confirmed also by the results of [31]. Here, during investigations of Raman scattering in CO_2 at a constant temperature, preservation of the line breadth of the fully symmetric vibration ν_1 during the gas—liquid transition was qualitatively observed.

CHAPTER V

INVESTIGATION OF THE RAMAN
SCATTERING SPECTRUM OF METHANE
AT VARIOUS PRESSURES [52]

§1. Structure of the Vibrational Band and Collision Broadening Spectrum for a Complex Molecule

For complex molecules, the observed breadth of the vibrational lines can include both unresolved structure lines and broadening of the individual components due to intermolecular interaction. The structure of the vibrational band of a polyatomic molecule has a complex character and is composed of a substantial number of branches superimposed on each other. For three-dimensional molecules there is an additional Coriolis splitting, comparable in magnitude with the breadth of the band structure. In view of the sizable moment of inertia of a polyatomic molecule, the bands are usually crowded together and form a line of complex shape. Only in the case of certain simple molecules (methane, for example) does the vibrational band possess sufficient separation between components to permit distinct observation of its structure. The structural regularities of the vibrational bands of complex molecules are considered in [53]. Calculation of the structure is possible in practice only for simple molecules, in particular for methane.

The breadth of the band which forms the structure is closely related to the type of vibration considered. For fully symmetric vibrations the bands are narrow, and the bands of vibrations which are not fully symmetric have a substantially greater size. The observed breadth of a band is determined by the intensities of its components. Since the intensity of a component is related to the population of the levels, the breadth Δ of a band depends on the temperature, as has been shown by L. L. Sobel'man [3]

$$\Delta \sim \text{const} \cdot T. \tag{V. 1}$$

For broadening of vibrational lines due to intermolecular interaction, it follows to expect the greatest effect from perturbation of the rotational motion during collisions, i.e., a change of the rotational state and a reorientation of the molecule. At room temperature each gas-kinetic collision should be accompanied by processes of this nature. The effect of perturbation of the rotational motion on different spectral lines is closely related to the nature of the lines considered and is determined [3] by the spatial anisotrophy of the tensor derivative of the polarizability $\partial\alpha_{ik}/\partial q_i$, which corresponds to the vibration along a normal coordinate. The relation between the isotropic and anisotropic parts of the tensor $\partial\alpha_{ik}/\partial q_i$ also determines the degree of depolarization ρ of the line, which can then be used as a convenient parameter in treating the broadening of the vibrational lines which arises from the perturbing of the rotation. Investigation shows that when the tensor $\partial\alpha_{ik}/\partial q_i$ is isotropic ($\rho = 0$, the line of a fully symmetric vibration) rotational motion is not reflected in the scattering amplitude, and broadening due to perturbation of the rotation vanishes. If the tensor is anisotropic ($\rho = {}^6/_7$, depolarized lines) the vibrational band is composed of transitions with a change of rotational state, and the perturbation of the rotation leads to broadening of the band components. In the intermediate case ($0 < \rho < {}^6/_7$) the role of perturbation of the rotation in broadening is determined by the relation between the isotropic and anisotropic parts of the tensor $\partial\alpha_{ik}/\partial q_i$. The difference in the line broadening, depending on their polarization, should also be observed in liquids, since the reorientation of the molecules during their rotationally chaotic motion is sufficient to preserve the mechanism considered.

TABLE 4. Fundamental Frequencies of Methane (1-0 transitions, gas) Which Appear in Raman Scattering

Assignment	ν, cm^{-1}, [56, 57]	I, rel. units [58]	ρ [58]
ν_1 (a_1)	2916.5	400	0
ν_2 (e)	1533.6	80	Depol.
ν_3 (f_2)	3018.7	400	Depol.

The broadening due to perturbation of the rotational motion could be detected qualitatively during investigation of the liquid—crystal phase transformation. In view of the change in character of the molecular motion there, it follows to expect a sharp change of the breadth of the depolarized lines. In addition, there is also possibly a change of the structure of the molecular band. Experimental investigations show that during transformation into the crystalline state a substantial narrowing of the depolarized lines is observed, and the change in the breadth of the polarized lines is comparatively small [54].

During investigations in liquids at different temperatures there will be a change of breadth due to perturbation of the rotational state and simultaneously a change in the breadth of the band structure. Investigations of the temperature dependence of the line breadth in Raman scattering in liquid isopentane [55], carried out by the author, show that for the isopentane lines a general narrowing is observed with a decreasing temperature. The relation between breadth and temperature has a nonlinear character. For polarized lines, the breadth and its change with temperature are comparatively small. For the depolarized lines, a substantial absolute value of the breadth is observed and it changes sharply with temperature. How much of the narrow broadening of the vibrational lines is connected with rotational motion can be established in investigations of the breadth in the gas at constant temperature. Since in this case the structure of the band does not change, the change of the contour with a change of pressure is entirely due to perturbation of the rotational states. Investigations of gases were not carried out until the present work. In our work, investigations of the line breadth of various vibrations were carried out on methane.

§2. Structure of the Vibrational Band in Raman Scattering in Methane

Methane is a tetrahedral molecule of the spherical top type. With this symmetry the rotational band ($\Delta v = 0$) is absent. Methane possesses four vibrations, of which three (Table 4) are observed in Raman scattering. In addition, the frequency $2\nu_2 = 3071$ cm^{-1} is observed in the Raman scattering spectrum, due to a Fermi resonance overtone $2\nu_2$ with the ν_3 frequency.

The ν_1 band corresponds to the fully symmetric vibration. According to the selection rules its rotational wings vanish, and the Q branch appears as a sharp intense line. Such a line can be observed if the difference between the rotational constants in the ground and the vibrational states is small: $B_e \approx B_v$. There are no data in the literature on the magnitude of the rotational constant for ν_1.

In the doubly degenerate ν_2 vibrational band, the Q branch and the O, P, R, and S rotational branches are observed. Since, for the ν_2 vibration, the moment of inertia is less in the excited state then in the ground state, the Q branch has a structure shadowed on the red side. In composition, its structure resembles one of the ν_3 Q branches (Fig. 24). At 300°K it forms a band whose breadth Δ_Q is about 7 cm^{-1}. For lines of the ν_2 vibrational band the Coriolis splitting vanishes.

In the triply degenerate ν_3 vibrational band of methane all five branches (Q, O, P, R, and S) are observed. Because of the term splitting in the higher vibrational states, due to the Coriolis interaction, each branch in the band is composed of three closely spaced Q^-, Q^0, and Q^+ branches. The rotational wings of the band present a complex picture of mutually superimposed lines. The structure of the Q branch at 300°K, calculated with the rotational constant for the ν_3 vibration [57], is shown in Fig. 24. The breadth of the band which forms the structure is about 12 cm^{-1}.

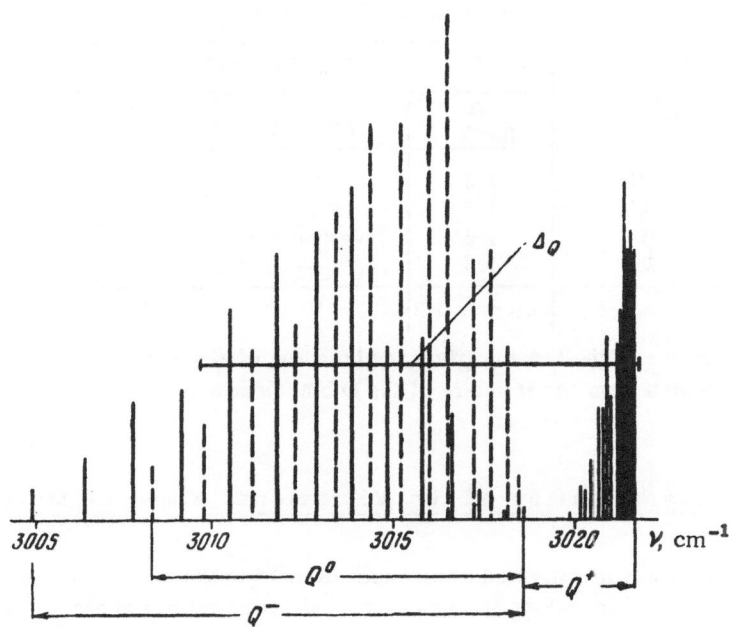

Fig. 24. Structure of the Q branch of the ν_3 vibrational band of methane at 300°K.

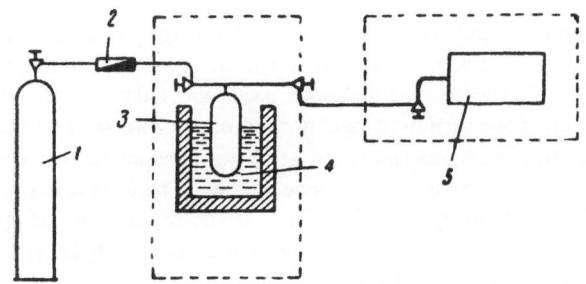

Fig. 25. Diagram of the apparatus for producing methane under high pressure. 1) Methane cylinder (50 atm); 2) filter; 3) high-pressure cylinder (1500 atm); 4) insulating vessel of foam plastic with liquid nitrogen; 5) illuminator for investigation of Raman scattering at high pressure (400 atm).

Thus a complete set of the different vibrations is observed in the methane spectrum, and their shapes can be investigated experimentally. On the basis of the values of the line depolarization ρ given in Table 4, it follows to expect preservation of the breadth for the Q branch of the ν_1 vibration and a broadening of the ν_2 and ν_3 lines with changing pressure. Since the ν_2 and ν_3 bands possess a structure with substantial breadth, the broadening begins to appear only at high pressure. The broadening of the ν_2 and ν_3 lines begins to be observed when the breadth γ_p of the components due to collision broadening exceeds the breadth of the band structure ($\gamma_p > \Delta_Q$). This occurs, at room temperature, for the ν_2 Q branch with $\gamma_p > 7$ cm^{-1}, and for the ν_3 Q branch with $\gamma_p > 12$ cm^{-1}. The given data show that the observed broadening of these lines can be expected only for substantial pressures.

TABLE 5. Change of the Raman Scattering Line Breadths
of Methane with Pressure

Pressure, atm	Line breadth, cm^{-1}		
	ν_1 ($\rho=0$)	ν_2 (depol.)	ν_3 (depol.)
15	1.5	7	12
40	1.5	7	12
100	1.5	Washed out	12
250	1.5	—	25
	Less than 1.5	7	12

Note: The last row gives the breadths of the bands
which form the structure of the Q branch at 300°K.

§3. Experimental Investigation of the Vibrational Bands of Methane at Various Pressures

The methane spectrum was investigated at pressures from 15 to 250 atm. With pressures up to 200 atm, the 400 atm illuminator was used for taking the spectrum, and with pressures above 200 atm, the 1500 atm illuminator. To obtain high pressures, a special apparatus was rigged up for liquification of methane by nitrogen (Fig. 25). The apparatus allowed us to produce and maintain methane under pressures up to 1500 atm. The conditions for photographing and processing the spectra were analogous to the conditions for photographing O_2 and N_2, described above.

On the photographs obtained (Fig. 26), the full Raman scattering spectrum of methane was observed; the intense ν_1 vibration line at 2916 cm^{-1}, and the weaker ν_2, ν_3, and $2\nu_2$ vibrational bands at 1534 cm^{-1}, 3019 cm^{-1}, and 3071 cm^{-1} respectively. In the ν_2 and ν_3 bands there was only a Q branch; the low-intensity rotational wings of these bands sank into the continuous background of the excitation spectrum. The shape and breadth of the vibrational lines at low pressures (up to 40 atm) corresponds to the structure considered above (§2) for the Q branches of the vibrational bands of methane. The ν_1 band is a sharp intense line of symmetric shape and breadth 1.5 ± 0.6 cm^{-1}. The ν_2 and ν_3 bands repeat the breadth and shape of the structure described earlier. The observed breadth and shape of the Q branch contours of the vibrational bands of methane are related to their structure, in agreement with [47].

The change of pressure starts to affect the line shape with pressures above 40 atm, and is reflected somewhat differently in the lines of different vibrations. The breadths of the methane lines as a function of pressure are given in Table 5. It can be seen from the table that the ν_1 band of the fully symmetric vibration ($\rho = 0$) does not change over the whole investigated region. In contrast to the ν_1 band, the Q branches of the ν_2 and ν_3 bands (both depolarized lines) respond to a change of pressure. These bands possess a broad structure; the change of breadth with increasing pressure begins when the breadth γ_p of the components overlaps the whole structure Δ_Q of the band. The Q branch of the ν_2 band has the same breadth for pressures up to 40 atm, but with greater pressures it begins to broaden and at 100 atm it dissolves into the background. The broadening of the Q branch of the ν_3 band begins with pressures above 100 atm; at 250 atm it is strongly smeared and its breadth amounts to about 25 cm^{-1}. The limiting pressure above which the broadening of the ν_2 and ν_3 lines begins to be observed ($\gamma_p = \Delta_Q$), and the data on breadth at higher pressures allow an estimate of the relative broadening γ_p/p of these lines. For the ν_2 Q branch this estimate gives $\gamma_p/p = 0.17$ cm^{-1}/atm and for the ν_3 Q branch, $\gamma_p/p \approx 0.1$ cm^{-1}/atm.

According to the collision theory, the relative broadening is determined by expression (III.6). Calculation by this formula shows that the effective optical collision radius for the broadening of the methane lines by intrinsic pressure is found to be 4-5 A for both the ν_2 and ν_3 lines. The value of ρ_ν obtained is of the same order as the analogous gas-kinetic diameter of methane, $d_0 = 4.1$ A.

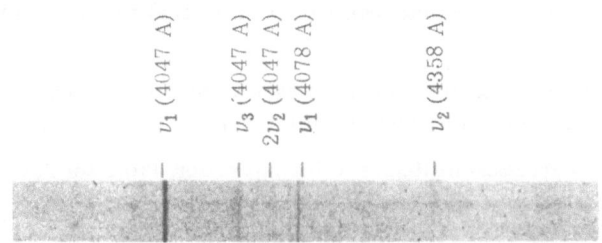

Fig. 26. Observed Raman scattering spectrum of methane
at 40 atm pressure.

§4. Discussion of the Results

Experimental investigation of the different vibrational lines of methane shows that their structure introduces a substantial contribution to the observed breadth (due to the vibration—rotation and Coriolis interactions). It should be noted that for polarized lines the breadth of the Q branch structure is substantially less than for depolarized lines.

The investigation revealed the different character of the collision broadening of the methane lines, depending on their polarization: for the polarized ν_1 line the collision broadening vanishes and for both of the depolarized ν_2 and ν_3 lines a substantial change of breadth with pressure is observed. The ν_1 line corresponds to a fully symmetric vibration and belongs to the same group of transitions as the previously investigated Q branches of O_2 and N_2; the $\partial \alpha_{ik}/\partial q_i$ tensor for ν_1 is isotropic. The absence of collision broadening for ν_1 again shows that for these transitions no perturbation of the rotational states of the molecule appears during collisions.

The substantial value of collision broadening for the depolarized ν_2 and ν_3 lines ($\rho_\nu \approx d_0$) allows us to assume that at the basis of their broadening lies a change of the rotational state during collisions. Proceeding from the mechanism whereby the rotational motion affects lines of different polarization, considered in the beginning of this chapter, it follows to expect a substantial broadening for these lines due to perturbation of the rotational motion. Since a change of rotational state occurs almost every time there is a gas-kinetic collision, it follows to expect a substantial effective cross section ($\rho_\nu \sim d_0$) for the broadening thus produced. This agrees with what is observed experimentally. For the Q branches $\Delta J = 0$, and the possible change of the rotational state consequently should be considered as a reorientation of the molecule, i.e., $\Delta m \neq 0$.

The investigation of methane shows that the collision broadening depends fundamentally on the type of transition considered. Analysis of the $2\nu_2$ band of methane in the infrared region [59] shows that here, too, a substantial collision broadening of the Q branch is observed with a change of the intrinsic pressure of the absorbing gas. For broadening by admixed gases it was found that the amount of collision broadening depends fundamentally on the nature of the perturbing gas and with the same gas it could appear differently in the Q branch and the rotational wings.

Investigations in a gas, where free rotation exists, show the significant role of the change of rotational state in the broadening of vibrational lines, and the close connection between the observed broadening and the polarization.

This supports the assumptions of L. L. Sobel'man about the mechanism of line broadening in liquids, connected with the rotational motion of the molecules. The works considered earlier [54, 55] and the later investigations of the line breadth in liquids [60] indicate the different character of the line broadening, depending on the polarization. On the basis of the present work it can be considered that the observed broadening in liquids is closely related to the rotational motion of the molecules.

It has been shown in [61] that for the temperature dependence of the line breadth in infrared absorption in toluene, depending on the spatial anisotropy of the tensor derivative $\partial M_{ik}/\partial q_i$ of the dipole moment M of the molecule along a normal coordinate, a different character of the absorption line broadening is observed. This indicates that an analogous relation between broadening and rotational motion of the molecules also exists in the infrared region.

The present work was carried out at first under the supervision of G. S. Landsberg. His constant concern and tactful advice will forever remain in the memory of the author.

In conclusion, the author expresses his thanks to his supervisor, Professor Pavel Alekseevich Bazhulin, for constant advice and aid in the work.

LITERATURE CITED

1. Sh. Chen and M. Takeo, Usp. Fiz. Nauk 66(3):391 (1958).
2. I. L. Sobel'man, Usp. Fiz. Nauk 54(4):551 (1954).
3. L. L. Sobel'man, Tr. Fiz. Inst. Akad. Nauk 9:316 (1958).
4. G. Kortüm and H. Verleger, Proc. Phys. Soc. 63:462 (1950).
5. H. Margenau, Rev. Mod. Phys. 11:1 (1939).
6. M. Mizushima, Phys. Rev. 83:94 (1951); 84:363 (1951).
7. N. Verleger, Z. Physik. 38:83 (1937).
8. E. Lindholm, Z. Physik. 109:223 (1938); 113:596 (1939).
9. D. Weber and S. Penner, J. Chem. Phys. 21:1503 (1958).
10. R. Coulon, B. Oksengorn, and St. Robin, Compt. Rend. 236:1481 (1953); R. Coulon, L. Galatry, B. Oksengorn, St. Robin, and B. Vodar, J. Phys. Radium, 15:58 (1954).
11. W. Benesh and T. Elder, Phys. Rev. 91:308 (1958).
12. W. V. Smith and R. Noward, Phys. Rev. 79:132 (1950).
13. R. M. Hill and W. V. Smith, Phys. Rev. 82:451 (1951).
14. F. Bruin, Thesis, Amsterdam (1956).
15. C. H. Townes, Phys. Rev. 70:665 (1946); B. Bleany and R. P. Penrose, Proc. Phys. Soc. 59:418 (1947); B. Bleany and J. H. N. Loubster, Nature 161:552 (1948), Proc. Phys. Soc. A63:483 (1950); D. E. Smith, Phys. Rev. 74:506 (1948); L. R. Weingarten, Thesis, Columbia Univ. (1948).
16. R. S. Anderson, W. V. Smith, and W. Gordy, Phys. Rev. 82:264 (1951); R. Beringer, Phys. Rev. 70:53 (1946); J. H. Van Vleck, Phys. Rev. 71:413 (1947); M. W. P. Strandberg, C. J. Meng, and L. G. Ingersoll, Phys. Rev. 75:1524 (1949); L. R. L. Lamont, Phys. Rev. 74:353 (1948); L. D. Artman and L. P. Gordon, Phys. Rev. 87:277 (1957).
17. G. E. Becker and S. H. Autler, Phys. Rev. 70:300 (1946).
18. G. L. Townes and F. R. Merritt, Phys. Rev. 70:558 (1946).
19. O. R. Gillam, H. D. Edwards, and W. Gordy, Phys. Rev. 75:1014 (1949).
20. G. L. Townes, R. Holden, and F. R. Merritt, Phys. Rev. 74:1113 (1948); C. M. Johnson and D. H. Slager, Phys. Rev. 87:667 (1952).
21. H. Feeney, H. A. Lackner, P. Moser, and W. V. Smith, J. Chem. Phys. 22:79 (1954).
22. A. G. Smith, W. Gordy, T. W. Simons, and W. V. Smith, Phys. Rev. 75:260 (1949).
23. R. Trambarulo, H. Lackner, P. Moser, and H. Feeney, Phys. Rev. 95:622(A) (1954).
24. R. S. Anderson, W. V. Smith, and W. Gordy, Phys. Rev. 87:561 (1952).
25. B. V. Gokhale and M. W. P. Strandberg, Phys. Rev. 84:884 (1951).
26. W. V. Smith, H. A. Lackner, and A. B. Volkor, J. Chem. Phys. 23:389 (1955).
27. P. W. Anderson, Phys. Rev. 76:647 (1949).
28. J. A. Van Vleck and V. F. Weisskopf, Rev. Mod. Phys. 17:227 (1945).
29. B. Bleany and R. P. Penrose, Proc. Phys. Soc. 60:540 (1948).
30. C. Fuchtlaner and W. Hoffman, Ann. Physik 43:96 (1914).
31. H. L. Welsh, P. E. Pasher, and B. P. Stoicheff, Can. J. Phys. 30:99 (1952).
32. Kh. E. Sterin, Candidate's dissertation, Tr. Fiz. Inst. Akad. Nauk 5:16 (1956).
33. Kh. E. Sterin, Izv. Akad. Nauk SSSR, Ser. Fiz. 14:411 (1950).
34. L. L. Sobel'man, Izv. Akad. Nauk SSSR, Ser. Fiz. 17:554 (1953).

35. F. Matossi, Phys. Rev. 76:1845 (1948).

36. M. F. Crawford, L. Welsh, and J. H. Harrold, Can. J. Phys. 30:81 (1952).

37. S. G. Rautian, Zh. Eksperim. i Teor. Fiz. 27:625 (1954).

38. M. L. Sosinskii, Izv. Akad. Nauk SSSR, Ser. Fiz. 17:621 (1953).

39. H. H. Glaassen and J. R. Nielsen, J. Opt. Soc. Am. 43:352 (1953).

40. H. L. Welsh, C. Cumming, and E. J. Stansbury, J. Opt. Soc. Am. 41:712 (1951).

41. J. H. Callomon, Can. J. Phys. 34:1046 (1956); H. L. Welsh, E. J. Stansbury, J. Romanko, and T. Feldman, J. Opt. Soc. Am. 45:338 (1955).

42. I. I. Breido, Usp. Fiz. Nauk 1:118 (1951).

43. G. B. Mikhailov, Zh. Eksperim. i Teor. Fiz. 37:1570 (1959).

44. G. Herzberg, Spectra and Structure of Diatomic Molecules [Russian translation], IL (1949), pp. 78, 338.

45. V. K. Ablekov, Opt. i Spektroskopiya 6:562 (1959).

46. S. G. Rautian, Usp. Fiz. Nauk 16(3):475 (1958).

47. P. A. Bazhulin and Yu. A. Lazarev, Opt. i Spektroskopiya 8:206 (1960).

48. G. V. Mikhailov, Zh. Eksperim. i Teor. Fiz. 36:1368 (1959).

49. M. F. Crawford, H. L. Welsh, and J. H. Harrold, Can. J. Phys. 30:81 (1952).

50. A. Loftus, Can. J. Phys. 35:216 (1957).

51. G. V. Mikhailov and A. V. Rakov, Apparatus and Test Stands, Theme 7, PS-55-467, Moscow (1955).

52. G. V. Mikhailov, Opt. i Sepktroskopiya 12: (5) (1962).

53. G. Herzberg, Vibrational and Rotational Spectra of Polyatomic Molecules [Russian translation], IL (1949).

54. A. V. Rakov, Proceedings of the Tenth All-Union Conference on Spectroscopy, Vol. I, L'vov (1957), p. 230.

55. G. V. Mikhailov, Proceedings of the Tenth All-Union Conference on Spectroscopy, Vol. I, L'vov (1957), p. 227.

56. T. Feldman, J. Romanko, and H. L. Welsh, Can. J. Phys. 33:138 (1955).

57. B. P. Stoicheff, C. Cumming, G. St. John, and H. L. Welsh, J. Chem. Phys. 20:498 (1952).

58. T. Yoshino and J. J. Bernstein, Conference on Molecular Spectroscopy, London (1958).

59. V. I. Malyshev, S. G. Rautian, and G. N. Zhizhin, Physical Problems of Spectroscopy, Izd. AN SSSR 2:14 (1963).

60. H. I. Resaev, Vestn. Mosk. Univ. (2):145 (1957).

61. A. V. Rakov, Present collection, p. 111.

BIBLIOGRAPHICAL INDEX OF PAPERS BY THE STAFF
OF THE OPTICAL LABORATORY

BIBLIOGRAPHICAL INDEX OF PAPERS
BY THE STAFF OF THE OPTICAL LABORATORY*

1934

Landsberg, G. S., Synopsis of Lectures on Optics, Lectures 1-37, Moscow, Dzerzh. Regional Council (1934), 165 pp; literature at the end of the lectures.

Landsberg, G. S., Joint Meeting of the Physical Institute, Moscow State University, and the Karpov Chemical Institute, Scientific and Technical Section (1934), No. 5-6, pp. 127-129.

Landsberg, G. S., Scattering of Light, Technical Encyclopedia (1934), Vol. 19, pp. 129-140; bibliography: 7 titles.

1935

Landsberg, G. S., Course of Optics, No. 1 (Lectures 1-13), Moscow, Moscow Correspondence University (1935), Vol. V, 196 pp., with figures.

Landsberg, G. S., Course of Optics, No. 2 (Lectures 14-23), Moscow, Department of Correspondence Instruction (1935), Vol. III, 118 pp.

Landsberg, G. S., Course of Optics, No. 3 (Lectures 24-38), Moscow, Department of Correspondence Instruction (1935), Vol. V, 148 pp., with figures.

Landsberg, G. S., Program of a General Course in Physics, Part 3, Optics, Moscow, Moscow State University, Physics Faculty (1935), 4 pp.

Landsberg, G. S., and Mandelstam, L. I., Selective light scattering in mercury vapor, in: Summaries of Reports to the Physics Group of the May Session of the Academy of Sciences, USSR; Leningrad, Izd. AN SSSR (1935), pp. 3-4.

Landsberg, G. S., Mandelstam, L. I., Tulyankin, S. V., and Zeiden, V. V., Spectral method of sorting alloy steels, Zavodsk. Lab. 4(10):1220-1227 (1935).

Landsberg, G. S., and Mandelstam, L. I., Selektive Lichtstreuung in Quecksilberdampf (Selective light scattering in mercury vapor), Soviet Phys. 8(4):378-400 (1935).

Landsberg, G. S., Mandelstam, L. I., Tulyankin, S. V., and Zeiden, V. V., Spektroskopische Methode zur Bestimmung von legierten Stälen (Spectroscopic methods of determining alloy steels), Techn. Phys. USSR 2(6):574-585 (1935).

1936

Landsberg, G. S., Contribution to discussion on a paper of D. S. Rozhdestvenskii, Analysis of spectra and spectral analysis, at the Session of the Academy of Sciences, USSR, March 14-20, 1936, Izv. Akad. Nauk SSSR, Ser. Fiz. (1-2):269-273 (1936).

---

*Complied by Senior Bibliographer of the FIAN Library A. A. Sakova. Edited by Senior Scientific Contributor V. I. Malyshev.

Landsberg, G. S., Electricity and Foundations of Molecular Physics, Moscow, Moscow State University, Physics Faculty (1936), 3 pp.

Landsberg, G. S., and Malyshev, V.I., Second-order lines in the Raman effect of light, Dokl. Akad. Nauk 3(8):365-367 (1936), bibliography: 5 titles; the same in French: Compt. Rend. Acad. Sci. URSS, 3(8):365-368.

Landsberg, G. S., and Mandelstam, L. I., Selektive Lichtstreuung in Quecksilberdampf (Selective light scattering in mercury vapor), Acta Phys. Polon. 5:79-84 (1936).

Landsberg, G. S., and Shubin, A. A., Rotating spark gap as stable source of intense ultraviolet light, Zh. Eksperim.i Teor. Fiz. 6(1):52-56 (1936).

Shubin, A. A., and Landsberg, G. S., On one simple method of photographic photometry, Zh. Eksperim. i Teor. Fiz. 1:57-59 (1936).

1937

Fabelinskii, L L., On the intensity ratio of certain lines in the mercury spectrum, Zh. Eksperim. i Teor. Fiz. 7(6):742-749 (1937); the same in German: Soviet Phys. 11(4):390-403 (1937).

Landsberg, G. S., Contribution to the discussion of a report to the Ural Petroleum Board, Work of the Ural physicotechnical institute, at the Meeting of the Physics Group, Academy of Sciences, USSR devoted to Questions of Optical Techniques, Izv. Akad. Nauk SSSR, Ser. Fiz. (6):887-889 (1937).

Landsberg, G. S., Raman effect of light, Physics Dictionary (1937), Vol. 3, Col. 30-41, bibliography: 6 titles.

Landsberg, G. S., On the precision of methods of works control, Zavodsk. Lab. 6(6):664-665 (1937).

Landsberg, G. S., Ways of developing spectral analysis, Izv. Akad. Nauk SSSR, Ser. Fiz. (2):101-112 (1937), bibliography: 4 titles.

Landsberg, G. S., Development of physical methods of testing in the USSR, Zavodsk. Lab. 6(11):1313-1320 (1937).

Landsberg, G. S., State of physics teaching in the higher technical schools (Report to Meeting of the Physical Group, Academy of Sciences, USSR on Questions of Teaching Physics in the Higher Technical Schools), Izv. Akad. Nauk SSSR, Ser. Fiz. (1):9-20 (1937).

Landsberg, G. S., Physics in the higher technical schools; Za. Prom. Kadry (2):34-40 (1937).

Landsberg, G. S., Physical methods of control of production and their significance for technology, Vestn. Akad. Nauk SSSR (7-8):45-48 (1937).

Landsberg, G. S., and Ukholin, S. A., Oscillation frequency of the hydroxyl group in methyl alcohol and its dependence on density, Dokl. Akad. Nauk 16(8):399-400 (1937); the same in French: Compt. Rend. Acad. Sci. URSS 16(8):391-393 (1937).

1938

Landsberg, G. S., Raman effect of light, Large Soviet Encyclopedia (1938), Vol. 33, pp. 562-566, bibliography: 6 titles.

Landsberg, G. S., Intermolecular forces and Raman effect of light, Izv. Akad. Nauk SSSR, Ser. Fiz. (3):373-382 (1938).

Landsberg, G. S., Heinrich Anton Lorentz (1853-1928), Large Soviet Encyclopedia (1938), Vol. 37, pp. 411-414, bibliography:10 titles.

Landsberg, G. S., and Malyshev, V. I., Raman spectra of a solution of water in dioxane and pyridine, Dokl. Akad. Nauk SSSR 18(8):549-551 (1938), bibliography: 5 titles; the same in French: Compt. Rend. Acad. Sci URSS 18(8):549-551 (1938).

1939

Bazhulin, P. A., The absorption of ultrasonic waves by electrolytes, J. Phys. USSR 1(5-6):431-437 (1939), bibliography: 10 titles.

Landsberg, G. S., Moscow conference on spectroscopy, Zavodsk. Lab. 8(8):777 (1939).

Landsberg, G. S., Modern optical methods of analytical chemistry, in: All-Union Conference on Analytical Chemistry, Moscow, November, 1939, Vol. 1; Moscow-Leningrad, Izd. AN SSSR (1939), pp. 89-113.

Landsberg, G. S., Malyshev, V. I., and Solov'ev, V. E., Raman effect of solutions of heavy water, Dokl. Akad. Nauk SSSR 24(9):873-874 (1939); the same in French: Compt. Rend. Acad. Sci. URSS 24(9):872-873 (1939).

Landsberg, G. S., and Shubin A. A., Light scattering in a nonuniformly heated crystal, Zh. Eksperim. i Teor. Fiz. 9(11):1309-1313 (1939), bibliography: 11 titles; the same in French: J. Phys. USSR 1(5-6):403-410 (1939).

1940

Fabelinskii, L. L., Depolarization of scattered light in mixtures of liquid stannic tetrachloride and tetrabromide, Izv. Akad. Nauk SSSR, Ser. Fiz. 4(1):166-168 (1940).

Fabelinskii, L. L., Die Depolisation von Zerstreutem Licht in Mischungen (Depolarization of scattered light in mixtures), J. Phys. USSR 2(4):277-282 (1940); bibliography: 5 titles.

Landsberg, G. S., Optics, Moscow-Leningrad, GTTI (1940) (General Course of Physics, Vol. 3), 559 pp., with figures.

Landsberg, G. S., Bazhulin, P. A., Rosenberg, Yu. V., and Eliner, A. S., Application of the Raman effect to the analysis of motor fuels, Izv. Akad. Nauk SSSR, Ser. Fiz. 4(1):158-160 (1940).

Landsberg, G. S., and Shubin, A. A., Heat conductivity of crystals, Zh. Eksperim. i Teor. Fiz. 10(2):247-249 (1940).

Malyshev, V. L., Intermolecular forces and Raman effect, Izv. Akad. Nauk SSSR, Ser. Fiz. 4(1):106-109 (1940).

Sterin, Kh. E., and Sushchinskii, M. M., Variation of the coefficient of absorption of oxygen with temperature in the extreme ultraviolet, Collection of Science Students' Papers, Moscow State University (1940), No. 14, pp. 15-38, bibliography: 14 titles.

Sterin, Kh. E., and Sushchinskii, M. M., Temperature dependence of the absorption of ultraviolet light by oxygen, Izv. Akad. Nauk SSSR, Ser. Fiz. 4(1):163-165 (1940).

1941

Bazhulin, P. A., Absorption of ultraacoustic waves in viscous liquids, Dokl. Akad. Nauk SSSR 31(2):114-117 (1941), bibliography: 8 titles.

Bazhulin, P. A., Spectral analysis of organic compounds by Raman effect methods, Izv. Akad. Nauk SSSR, Ser. Fiz. 5(2-3):168-173, bibliography: 3 titles.

Bazhulin, P. A., Platé, A. F., Solovova, O. P., and Kazanskii, B. A., Optical methods of studying hydrocarbons, II. Raman spectra of paraffins, Izv. Akad. Nauk SSSR, Otd. Khim. Nauk (1):13-26 (1941), bibliography: 6 titles.

Kazanskii, B. A., Solovova, O. P., and Bazhulin, P. A., Hydrolyzation of homologs of cyclopentane with splitting of cycles, Izv. Akad. Nauk SSSR, Otd. Khim. Nauk (1):107-114 (1941), bibliography: 11 titles.

Landsberg, G. S., All-Union Conference on Spectroscopy, Moscow, December 9-14, 1940; Vestn. Akad. Nauk SSSR (2-3):97-103 (1941).

Landsberg, G. S., Science and technology, 1940, Zavodsk. Lab. 10(1):11-12 (1941).

Landsberg, G. S., News on the investigation of processes in light sources and on the application of methods of spectral analysis, Zavodsk. Lab. 10(1):11-12 (1941).

Landsberg, G. S., Modern state of applied spectroscopy and its routine problems, Izv. Akad. Nauk SSSR, Ser. Fiz. 5(2-3):70-90 (1941), bibliography: 16 titles.

Landsberg, G. S., Spectral analysis, Zavodsk. Lab. 10(6):563-564 (1941).

Landsberg, G. S., Spectral analysis and its technical application, Sov. Nauka (2):51-62 (1941).

Malyshev, V. L., Study of intermolecular forces by the Raman effect method, Izv. Akad. Nauk SSSR, Ser. Fiz. 5(1):13-18 (1941).

Sushchinskii, M. M., Variation of Raman effect line intensities in solutions with concentration, Dokl. Akad. Nauk SSSR 33(1):21-24 (1941), bibliography: 3 titles.

Zelinskii, N. D., and Landsberg, G. S., Optical methods of studying hydrocarbons, I. Problems of the Raman spectroscopy of hydrocarbons and possibilities of their use, Izv. Akad. Nauk SSSR, Otd. Khim. Nauk (1):9-12 (1941).

1943

Bazhulin, P. A., Bokshtein, M. F., Liberman, A. L., Lukina, M. Yu., Magolis, E. L, Solovova, O. P., and Kazanskii, B. A., Optical methods of studying hydrocarbons, III. Raman spectra of paraffins, Izv. Akad. Nauk SSSR, Otd. Khim. Nauk (3):198-205 (1943).

1944

Bazhulin, P. A., Application of the Raman effect method to the analysis of hydrocarbon mixtures, Transactions of the All-Union Conference on Analytical Chemistry (1944), Vol. III, pp. 105-110.

Landsberg, G. S., Raman effect and its application in organic chemistry, in: Reports of the Committee on Hydrocarbons, No. 1. Optical Methods of Analyzing Hydrocarbon Mixtures, Moscow, Gostekhizdat (1944), pp. 3-50, bibliography: 3 titles.

1945

Bazhulin, P. A., Scattering spectra of individual hydrocarbons (Contribution to the Conference on Theoretical Spectroscopy, Moscow, 1944), Izv. Akad. Nauk SSSR, Ser. Fiz. 9(3):222-224 (1945).

Fabelinskii, I. L., Intensity distribution in theRayleigh line wing and relaxation phenomena in liquids (Contribution to the Conference on Theoretical Spectroscopy, Moscow, 1944), Izv. Akad. Nauk SSSR, Ser. Fiz. 9(3):186-191 (1945), bibliography: 15 titles.

Landsberg, G. S., Investigation of L. I. Mandelstam in the field of optics and molecular physics, Izv. Akad. Nauk SSSR, Ser. Fiz. 9(1-2):21-29 (1945).

Landsberg, G. S., State of spectral analysis in the USSR and measures proposed for its development (Contribution to the Conference on Spectral Analysis, Moscow, 1944), Izv. Akad. Nauk SSSR, Ser. Fiz. 9(6):585-592 (1945).

Landsberg, G., La diffusion de la lumière et les forces intramoléculaires (Diffusion of light and intermolecular forces), J. Phys. Radium 6(12):305-313 (1945), bibliography: 6 titles.

Landsberg, G. S., L. I. Mandelstam, 1879-1944, Zh. Eksperim. i Teor. Fiz. 15(1-2):3-6 (1945).

Malyshev, V. L, Spectroscopical methods of determining the heat of the hydrogen bond (Contribution to Conference on Theoretical Spectroscopy, Moscow, 1944), Izv. Akad. Nauk SSSR, Ser. Fiz. 9(3):198-200 (1945), bibliography: 2 titles.

1946

Bazhulin, P. A., Sterin, Kh. E., Bulanova, T. F., Solovova, O. P., Turova-Polyak, M. V., and Kazanskii, B. A., Optical method of studying hydrocarbons, Contribution IV. Raman spectra of naphthenes, Izv. Akad. Nauk SSSR, Otd. Khim. Nauk (1):7-18, bibliography: 17 titles.

Fabelinskii, L, Depolarization of scattered light in certain hydrocarbons of the paraffin series, Zh. Eksperim. i Teor. Fiz. 16(8):728-733 (1946), bibliography: 8 titles; the same in English: J. Phys. USSR 10(3):231-235 (1946).

Landsberg, G. La diffusion de la lumière et les forces intramoléculaires (Diffusion of light and intermolecular forces), in: Abstracts of Research Papers for 1945, Otd. Fiz.-Matem. Nauk, Moscow-Leningrad, Izd. AN SSSR (1946), p. 16.

Landsberg, G. S., and Baryshanskaya, F. S., Raman effect in crystalline hydroxides and the hydrogen bond, Izv. Akad. Nauk SSSR, Ser. Fiz. 10(5-6):509-522 (1946), bibliography: 27 titles; the same in English: J. Phys. USSR 11(2):101-111 (1947).

1947

Bazhulin, P. A., and Leontovich, M. A., Absorption of ultrasonic waves in liquids, Dokl. Akad. Nauk SSSR, 57(1):29-30 (1947), bibliography: 4 titles.

Bazhulin, P. A., and Sterin, Kh. E., Optical method of studying hydrocarbons, Raman spectra of the alkenes, Izv. Akad. Nauk SSSR, Ser. Fiz. 11(4):456-460 (1947), bibliography: 9 titles.

Fabelinskii, I. L., Rayleigh line wings in viscous liquids (Contribution to All-Union Conference on Spectroscopy, Leningrad, 1946), Izv. Akad. Nauk SSSR, Ser. Fiz. 11(4):382-389 (1947), bibliography: 11 titles.

Fabelinskii, I. L., Method of obtaining a continuous photometric wedge by cathode sputtering, Zh. Tekhn. Fiz. 17(9):1091-1096 (1947).

Fabelinskii, I. L., Effect of viscosity on the breadth of the Rayleigh line wing, Dokl. Akad. Nauk SSSR 57(4):341-342 (1947), bibliography: 5 titles.

Landsberg, G. S., Introductory remarks (at the opening of the All-Union Conference on Spectroscopy), Izv. Akad. Nauk SSSR, Ser. Fiz. 11(3):215-216 (1947).

Landsberg, G. S., Molecular spectral analysis and its development in the USSR, Zavodsk. Lab. 11(11):1290-1298 (1947).

Landsberg, G. S., Optics (2nd edn., revised), Moscow-Leningrad, Gostekhizdat (1947), 631 pp. (General Course on Physics, Vol. 3).

Landsberg, G. S., The field of Langevin (1872-1946), Usp. Fiz. Nauk 31(3):289-296 (1947).

Landsberg, G. S., and Baryshanskaya, F. S., Raman scattering of light in crystalline hydroxides and the hydrogen bond, Izv. Akad. Nauk SSSR, Ser. Fiz. 11(4):335 (1947).

Landsberg, G. S., and Baryshanskaya, F. S., Light scattering in crystalline hydroxides and the hydrogen bond, J. Phys. USSR 11(2):101-111 (1947).

Motulevich, G. P., Molecular light scattering in crystals, Izv. Akad. Nauk SSSR, Ser. Fiz. 11(4):390-400 (1947), bibliography: 21 titles.

Sushchinskii, M. M., Intensity and polarization of the Raman effect lines of the CH group in paraffins (Contribution to All-Union Conference on Spectroscopy, Leningrad, 1946), Izv. Akad. Nauk SSSR, Ser. Fiz. 11(4):341-344 (1947), bibliography: 4 titles.

Sushchinskii, M. M., Photometric method of determining the form and breadth of spectral lines (Contribution to the All-Union Conference on Spectroscopy, Leningrad, 1946), Izv. Akad. Nauk SSSR, Ser. Fiz. 11(4):348-352 (1947), bibliography: 6 titles.

1948

Baryshanskaya, F. S., Landsberg, G. S., and Motulevich, G. P., Scattering of light in ferroelectrics and the hydrogen bond, Izv. Akad. Nauk SSSR, Ser. Fiz. 12(5):573-575 (1948), bibliography: 8 titles.

Khaikin, S. E., and Landsberg, G. S., Leonid Isakovich Mandelstam (1879-1944), in: People of Russian Science, Vol. 1, Moscow-Leningrad, Gostekhizdat (1948), pp. 260-271.

Landsberg, G. S., Introductory remarks (at the opening of the Sixth Conference on Spectroscopy, Izv. Akad. Nauk SSSR, Ser. Fiz. 12(4):355-357 (1948).

Landsberg, G. S., Sixth Conference on Spectroscopy, Vestn. Akad. Nauk SSSR (11):74-76 (1948).

Landsberg, G. S., Doppler-Fizeau effect and molecular motion, Usp. Fiz. Nauk 36(3):284-291 (1948), bibliography: 12 titles.

Landsberg, G. S., and Baryshanskaya, F. S., Light scattering in KH_2PO_4 and $(NH_4)H_2PO_4$ and its significance for the theory of ferroelectrics, Dokl. Akad. Nauk SSSR 61(6):1027-1030 (1948), bibliography: 11 titles.

Sushchinskii, M. M., Basis of quantitative molecular analysis by the Raman effect method (Contribution to the Sixth Conference on Spectroscopy, Kiev, 1948), Izv. Akad. Nauk SSSR, Ser. Fiz. 12(5):567-572 (1948), bibliography: 9 titles.

Sushchinskii, M. M., Basis of quantitative molecular analysis by the Raman effect method, Zavodsk. Lab. 14(9):1070-1079 (1948), bibliography: 7 titles.

1949

Landsberg, G. S., Photography and spectroscopy, in: Summaries of Contributions to the Conference on Research Applications of Photography and Cinematography, December 14-16, 1949; Moscow-Leningrad, Izd. AN SSSR (1949), pp. 6-7.

Landsberg, G. S., Kazanskii, B. A., et al., Complex Method of a Detailed Study of the Individual Composition of Gasolines, Moscow, Gostekhizdat (1949), 68 pp., with figures; bibliography: 23 titles.

Malyshev, V. I., and Shishikina, M. V., Raman effect in the higher alcohols and the hydrogen bond, Dokl. Akad. Nauk SSSR 66(5):833-836 (1949), bibliography: 6 titles.

Mash, D., Mayants, L., and Fabelinskii, L, Measuring the temperature dependence of the dielectric constant and loss angle of dielectrics in the centimeter wave field, Zh. Tekhn. Fiz. 19(10):1192-1198 (1949), bibliography: 1 title.

1950

Bazhulin, P. A., Absorption of ultraacoustic waves in liquids (Dissertation in pursuance of the degree of Doctor of Physicomathematical Science), Tr. Fiz. Inst. Akad. Nauk 5:261-338 (1950), bibliography: 76 titles.

Landsberg, G. S., Introductory remarks to the Seventh All-Union Conference on Spectroscopy, Sverdlovsk, May 22-29, 1950, Izv. Akad. Nauk SSSR, Ser. Fiz. 14(4):383-386 (1950).

Malyshev, V. L, and Shishikina, M. V., Study in the association of a series of saturated monoatomic alcohols by the Raman effect method, Zh. Eksperim. i Teor. Fiz. 20(4):297-303 (1950), bibliography: 6 titles.

Malyshev, V. L, Methods of increasing the dispersion of spectral apparatus (Contribution to the Seventh All-Union Conference on Spectroscopy, May 22-29, 1950), Izv. Akad. Nauk SSSR, Ser. Fiz. 14(6):746-752 (1950), bibliography: 6 titles.

Motulevich, G. P., Molecular light scattering in crystals (Dissertation in pursuance of the Degree of Doctor of Physicomathematical Science), Tr. Fiz. Inst. Akad. Nauk 5:9-62 (1950), bibliography: 31 titles.

Motulevich, G. P., and Fabelinskii, L L., On the Rayleigh scattering of light in crystals (Contribution to the Seventh All-Union Conference on Spectroscopy, May 22-29, 1950), Izv. Akad. Nauk SSSR, Ser. Fiz. 14(4):542-548 (1950).

Motulevich, G. P., Fabelinskii, I. L., and Steingaus, L. N., Absolute acoustic microradiometer, Dokl. Akad. Nauk SSSR 70(1):29-31 (1950).

Motulevich, G. P., and Turovtseva, Z. M., Molecular scattering in iceland spar, Zh. Eksperim. i Teor. Fiz. 20(4):334-339 (1950), bibliography: 11 titles.

Rytov, S. M., and Fabelinskii, L L., New form of phase diffraction grating, Zh. Eksperim. i Teor. Fiz. 20(4):340-341 (1950), bibliography: 4 titles.

Shubin, A. A., Infrared spectra of carboxylic acids (Contribution to the Seventh All-Union Conference on Spectroscopy, May 22-29, 1950), Izv. Akad. Nauk SSSR, Ser. Fiz. 14(4):442-451 (1950), bibliography: 4 titles.

Sushchinskii, M. M., Molecular analysis by the Raman effect method (Dissertation in pursuance of the Degree of Master of Physicomathematical Science), Tr. Fiz. Inst. Akad. Nauk 5:185-260 (1950), bibliography: 76 titles.

Sushchinskii, M. M., Automatic recording photoelectric spectrograph for studying Raman spectra, Dokl. Akad. Nauk SSSR 70(2):221-224 (1950), bibliography: 5 titles.

Sushchinskii, M. M., Photometric spectrophotometer for studying Raman spectra and its application for analytical purposes (Contribution to the Seventh All-Union Conference on Spectroscopy, May 22-29, 1950), Izv. Akad. Nauk SSSR, Ser. Fiz. 14(4):387-392 (1950), bibliography: 9 titles.

Sushchinskii, M. M., Photoelectric method of studying Raman spectra, Zh. Eksperim. i Teor. Fiz. (4):304-317 (1950), bibliography: 15 titles.

Velichkina, T. S., and Fabelinskii, L L., Method of measuring the velocity of propagation of ultrasonic waves in a liquid, Dokl. Akad. Nauk SSSR 75(2):177-180 (1950).

1951

Landsberg, G. S., Optical methods of molecular spectral analysis of organic liquids, in: Summaries of Contributions to the All-Union Conference on Chemistry and the Processing of Petroleum, Baku, AN Azerb. SSR (1951), pp. 7-9.

Landsberg, G. S., and Kazanskii, B. A., Determination of the individual composition of gasolines by direct distillation by the Raman method, Izv. Akad. Nauk SSSR, Otd. Khim. Nauk (2):100-114 (1951), bibliography: 17 titles.

Motulevich, G. P., and Fabelinskii, L L., On one optical method of controlling the character of the acoustic field, Dokl. Akad. Nauk SSSR, 81(5):787-790 (1951), bibliography: 2 titles.

1952

Landsberg, G. S., Optics, 3rd ed. revised, Moscow, Gostekhizdat (1952), 727 pp., with figures.

Landsberg, G. S., and Baryshanskaya, F. S., Effect of temperature on the scattering spectrum of crystals containing OH groups, in: In Memory of Sergei Ivanovich Vavilov, Moscow, Izd. AN SSSR (1952), pp. 147-156, bibliography: 14 titles.

Landsberg, G. S., and Zelinskii, N. D., Petr Petrovich Lazarev (On the Tenth Anniversary of his Death), Vestn. Akad. Nauk SSSR (4):67-69 (1952).

Malyshev, B. L., Markov, M. N., and Shubin, A. A., Automatic infrared spectrophotometer, Dokl. Akad. Nauk SSSR 86(2):273-276 (1952), bibliography: 5 titles.

Motulevich, G. P., and Fabelinskii, L. L., Molecular scattering of light in liquids, Dokl. Akad. Nauk SSSR, 83(2):203-206 (1952), bibliography: 12 titles.

Sushchinskii, M. M., Breadth of lines in Raman spectra of certain hydrocarbons, I, Zh. Eksperim. i Teor. Fiz. 22(6):755-767 (1952), bibliography: 16 titles.

1953

Bazhulin, P. A., Egorov, Yu. P., and Mironov, V. F., Optical method of studying silicoorganic compounds. Additivity of the intensities of the characteristic frequencies in the Raman spectra of allylsilanes, Dokl. Akad. Nauk SSSR 92(3):515-517 (1953), bibliography: 8 titles.

Bazhulin, P. A., and Peregudov, G. V., Improved method of molecular analysis by Raman spectra (Contribution to the Eighth All-Union Conference on Spectroscopy, Moscow, June 2-10, 1952), Izv. Akad. Nauk SSSR, Ser. Fiz. 17(5):617-620 (1953), bibliography: 4 titles.

Egorov, Yu. P., and Bazhulin, P. A., Optical method of studying silicoorganic hydrocarbons, Raman spectra of alkenylsilanes, Dokl. Akad. Nauk SSSR 88(4):647-650 (1953), bibliography: 10 titles.

Fabelinskii, L. L., Spectral study of the classical scattering of light in liquids (Contribution to the Eighth All-Union Conference on Spectroscopy, Moscow, 1952), Izv. Akad. Nauk SSSR, Ser. Fiz. 17(5):538-545, 552-553 (1953), bibliography: 12 titles.

Fabelinskii, L. L., and Shustin, O. A., Dispersion of the velocity of sound in certain organic liquids, Dokl. Akad. Nauk SSSR 92(2):285-288 (1953), bibliography: 10 titles.

Landsberg, G. S., The Raman effect of light, Large Soviet Encyclopedia, 2nd ed. (1953), Vol. 22, pp. 121-124, bibliography: 5 titles.

Landsberg, G. S., Langevin field. Large Soviet Encyclopedia, 2nd ed. (1953), Vol. 24, pp. 279-280; bibliography: 6 titles.

Landsberg, G. S., Methods of molecular spectral analysis of organic liquids, in: Transactions of the All-Union Conference on Chemistry and the Processing of Petroleum, September 18-24, 1951, Baku, AN Azerb. SSR (1953), pp. 7-20.

Landsberg, G. S., Optical methods of studying molecules, Priroda (1):19-26 (1953); (2):8-14 (1953).

Landsberg, G. S., Bazhulin, P. A., and Sushchinskii, M. M., Atlas of Raman spectra of individual hydrocarbons (Contribution to the Eighth All-Union Conference on Spectroscopy, Moscow, 1952), Izv. Akad. Nauk SSSR, Ser. Fiz. 17(5):604-606 (1953), bibliography: 3 titles.

Malyshev, V. L., Markov, M. N., and Shubin, A. A., On a two-beam infrared spectrophotometer, Izv. Akad. Nauk SSSR, Ser. Fiz. 17(5):654-659 (1953).

Motulevich, G. P., and Fabelinskii, L. L., Study of the Acoustic Radiation of Barium Titanate, Zh. Eksperim. i Teor. Fiz. 25(5):605-613 (1953), bibliography: 6 titles.

Sobel'man, L. L., Breadth of Raman effect lines in steam, Izv. Akad. Nauk SSSR, Ser. Fiz. 17(5):554-560 (1953).

Sobel'man, L. L., Rayleigh scattering line breadth in gas, Dokl. Akad. Nauk SSSR, 88(4):653-656 (1953), bibliography: 5 titles.

Sobel'man, L. L., and Vainshtein, L. A., Broadening of spectral lines as a result of the quadrupolar stark effect, Dokl. Akad. Nauk SSSR 90(5):757-760 (1953), bibliography: 10 titles.

Sushchinskii, M. M., Finding the true Raman effect line shape from the shape observed, Zh. Eksperim. i Teor. Fiz. 25(1):87-94 (1953), bibliography: 10 titles.

Sushchinskii, M. M., Characteristic lines in Raman effect spectra of hydrocarbons (Contribution to the Eighth All-Union Conference on Spectroscopy, Moscow, 1952), Izv. Akad. Nauk SSSR, Ser. Fiz. 17(5):608-616 (1953), bibliography: 24 titles.

Velichkina, T. S., Molecular scattering of light in viscous liquids and amorphous solids (Contribution to the Eighth All-Union Conference on Spectroscopy, Moscow, June 2-10, 1952), Izv. Akad. Nauk SSSR, Ser. Fiz. 17(5):546-553 (1953), bibliography: 11 titles.

1954

Bazhulin, P. A., Koperina, A. V., Liberman, A. L., Ovodova, V. A., and Kazanskii, B. A., Optical method of studying hydrocarbons, Raman spectra of certain naphthenes, Izv. Akad. Nauk SSSR, Otd. Khim. Nauk (4):709-715 (1954), bibliography: 16 titles.

Bazhulin, P. A., Rautian, S. G., Sokolovskaya, A. I., and Sushchinskii, M. M., Methods of studying Raman effect line breadths (Contribution to the Ninth All-Union Conference on Spectroscopy, Tartu, June 5-11, 1954), Izv. Akad. Nauk SSSR, Ser. Fiz. 18(6):678-679 (1954), bibliography: 3 titles.

Ginzburg, V. L., and Motulevich, G. P., Optical properties of metals (Contribution to the Ninth All-Union Conference of Spectroscopy, Tartu, June 5-11, 1954), Izv. Akad. Nauk SSSR, Ser. Fiz. 18(6):631-634 (1954), bibliography: 16 titles.

Kazanskii, B. A., Landsberg, G. S., Aleksanyan, V. T., Bulanova, T. F., Liberman, A. L., Mikhailova, E. A., Platé, A. F., Sterin, Kh. E., and Ukholin, S. A., Analysis of the aromatic part of ligroin from Raman spectra (Contribution to the Ninth All-Union Conference on Spectroscopy, Tartu, June 5-11, 1954), Izv. Akad. Nauk SSSR, Ser. Fiz. 18(6):704-706 (1954).

Kazanskii, B. A., Landsberg, G. S., Platé, A. F., Bazhulin, P. A., Liberman, A. L., Mikhailova, E. A., Sushchinskii, M. M., Tarasova, G. A., Ukholin, S. A., and Voron'ko, S. V., Determination of the individual hydrocarbon composition of gasolines by the Raman method, Communication 5. Gasoline from Emba petroleum, Izv. Akad. Nauk SSSR, Otd. Khim. Nauk (5):865-877 (1954), bibliography: 4 titles.

Kazanskii, B. A., Landsberg, G. S., Platé, A. F., Bazhulin, P. A., Mikhailova, E. A., Liberman, A. L., Sushchinskii, M. M., Tarasova, G. A., Ukholin, S. A., and Voron'ko, S. V., Determination of the individual hydrocarbon composition of gasolines by the Raman method. Communication 4. Gasoline from Tuimazy oil petroleum, Izv. Akad. Nauk SSSR, Otd. Khim. Nauk (3):456-469 (1954).

Kazanskii, B. A., Landsberg, G. S., Platé, A. F., Liberman, A. L., Mikhailova, E. A., Bazhulin, P. A., Batuev, M. I., Ukholin, S. A., Bulanova, T. F., and Tarasova, G. A., Determination of the individual hydrocarbon composition of gasolines by the Raman method, Communication 3. Surakhansk gasolines, Izv. Akad. Nauk SSSR, Otd. Khim. Nauk (2):278-291 (1954), bibliography: 2 titles.

Kazanskii, B. A., Landsberg, G. S., Platé, A. F., Liberman, A. L., Mikhailova, E. A., Sterin, Kh. E., Ulanova, T. F., Tarasova, G. A., and Aleksanyan, V. T., Determination of the individual hydrocarbon composition of gasoline by the Raman method, Communication 6. Karachukhursk gasoline, Izv. Akad. Nauk SSSR, Otd. Khim. Nauk (6):1053-1076 (1954), bibliography: 5 titles.

Landsberg, G. S., Molecular scattering of light and its application, Uli Tunbao (6):321-329 (1954) (in Chinese).

Landsberg, G. S., and Kazanskii, B. A., Regarding an article by M. D. Tilicheev, On the Soviet atlas of Raman spectra of hydrocarbons [Neft. Khoz. (1) (1953)]. Neft. Khoz. (3):31-36 (1954), bibliography: 15 titles.

Landsberg, G. S., Shatenshtein, A. I., Peregurov, G. V., Izrailevich, E. A., and Novikova, L. A., Vibration spectra of the diphenyl and decadeuterodiphenyl molecules, Izv. Akad. Nauk SSSR, Ser. Fiz. 18(6):669-671 (1954), bibliography: 5 titles.

Rautian, S. G., Choice of condenser in studying Raman effects, Zh. Eksperim. i Teor. Fiz. 27(5):625-635 (1954), bibliography: 13 titles.

Sobel'man, I. I., On the relation of the statistical and collision theory of the broadening of spectral lines, Dokl. Akad. Nauk SSSR 98(1):43-45 (1954), bibliography: 11 titles.

Sobel'man, I. I., On the theory of the breadth of atomic spectral lines, Usp. Fiz. Nauk 54(4):551-586 (1954), bibliography: 60 titles; the same in German: Fortschr. Physik 5(4):175-210 (1957).

Sushchinskii, M. M., and Tyulin, V. I., Study of the degree of depolarization of Raman lines of hydrocarbons with conjugate double bonds, Dokl. Akad. Nauk SSSR 95(3):505-508 (1954), bibliography: 21 titles.

1955

Bazhulin, P. A., and Rakov, A. V., Study of the breadth of Raman lines as a function of the state of aggregation of the substance, Dokl. Akad. Nauk SSSR 105(1):54-56 (1955), bibliography: 8 titles.

Bazhulin, P. A., Rautian, S. G., Sokolovskaya, A. L, and Sushchinskii, M. M., Methods of studying the breadth of Raman lines and their application, Zh. Eksperim. i Teor. Fiz. 29(6):822-829 (1955), bibliography: 17 titles.

Ginzburg, V. L., and Motulevich, G. P., Optical properties of metals, Usp. Fiz. Nauk 55(4):469-535 (1955), bibliography: 64 titles.

Landsberg, G. S., Optics, Large Soviet Encyclopedia, 2nd ed. (1955), Vol. 31, pp. 95-102, bibliography: 11 titles.

Landsberg, G. S., and Sushchinskii, M. M., Die Methode der Molekularspektroskopie und deren Anwendung für die Analyse der Erdölerzeugnisse [Methods of molecular spectroscopy and their application to the analysis of petroleum products (Contribution to the Conference on Spectroscopy, Physical Society of the German Democratic Republic, Weimar, 1955)], Exptl. Techn. Phys. 3 (separate part):106-114 (1955).

Molchanov, V. A., and Fabelinskii, L L., Dispersion of the velocity of sound in carbon disulfide, Dokl. Akad. Nauk SSSR 105(2):248-249 (1955), bibliography: 4 titles.

Sobel'man, L L, Applicability of the statistical treatment of the broadening of spectral lines as a result of pressure and the doppler effect (Contribution to the Ninth All-Union Conference on Spectroscopy, Tartu, 1954), Izv. Akad. Nauk SSSR, Ser. Fiz. (19):24-25 (1955), bibliography: 9 titles.

1956

Fabelinskii, L L., Intensity ratios in the fine structure components of the Rayleigh light scattering line, Dokl. Akad. Nauk SSSR, 106(5):822-825 (1956), bibliography: 17 titles.

Landsberg, G. S., Tenth All-Union Conference on Spectroscopy, L'vov, 1956; Priroda (11):43-47 (1956).

Landsberg, G. S., Sixth international colloquium on spectroscopy in Amsterdam, May, 1956; Usp. Fiz. Nauk 60(4):722-734 (1956).

Landsberg, G. S., Shatenshtein, A. L, Peregudov, G. V., Izrailevich, E. A., and Novikova, L. A., Vibration spectra of diphenyl and decadeuterodiphenyl, Opt. i Spektroskopiya 1(1):34-40 (1956).

Markov, M. N., Measurement of low-frequency current noise from a low internal resistance source, Pribory i Tekhn. Eksperim. (3):70-73 (1956), bibliography: 10 titles.

Markova, S. V., Bazhulin, P. A., and Sushchinskii, M. M., Optical method of studying hydrocarbons. Raman spectra of unsaturated hydrocarbons, Opt. i Spektroskopiya 1(1):41-53 (1956), bibliography: 22 titles.

Motulevich, G. P., and Fabelinskii, L L., Variation of the refractive index with density at low sonic frequencies, Dokl. Akad. Nauk SSSR, 106(4):637-640 (1956), bibliography: 5 titles.

Rautian, S. G., Effect of the curvature of the spectral line on the form of the transmission function of a monochromator, Opt. i Spektroskopiya 1(8):1000-1006 (1956), bibliography: 13 titles.

Rautian, S. G., Extent of the resolving power of optical apparatus, Dokl. Akad. Nauk SSSR 109(4):743-745 (1956), bibliography: 10 titles.

Rautian, S. G., Illumination of the spectrograph by a volume source, Opt. i Spektroskopiya 1(3):439-440 (1956), bibliography: 8 titles.

Rezaev, N. L, and Bazhulin, P. A., Measurement of the contours and breadths of Raman lines by photoelectric recording, Opt. i Spektroskopiya 1(5):715-718 (1956), bibliography: 8 titles.

Sobel'man, L L, Broadening of spectral lines owing to collisions with electrons, Zh. Eksperim. i Teor. Fiz. 31(3):519-520 (1956), bibliography: 11 titles; Opt. i Spektroskopiya 1(5):617-626 (1956), bibliography: 14 titles.

1957

Fabelinskii, L L., Depolarization scattering of light in liquids and relaxation processes, Opt. i Sepktroskopiya 2(4):510-513 (1957), bibliography: 13 titles.

Fabelinskii, L L., Wing of the Rayleigh line and relaxation processes in liquids (Contribution to the Tenth All-Union Conference on Spectroscopy, July 1956), Fiz. Sb. L'vovsh. Gos. Univ (3):117-118 (1957), bibliography: 6 titles.

Fabelinskii, L L., Some questions on the molecular scattering of light in liquids, Usp. Fiz. Nauk 63(2):355-410 (1957), bibliography: 122 titles.

Landsberg, G. S., Introductory remarks (to the Tenth All-Union Conference on Spectroscopy), in: Material of the Tenth All-Union Conference on Spectroscopy, Vol. 1, L'vov; Izd. L'vov Univ. (1957), pp. 7-13.

Malyshev, V. I., Study of the hydrogen bond by spectroscopical methods, Usp. Fiz. Nauk 63(2):323-353 (1957), bibliography: 43 titles.

Malyshev, V. I., Study of the transmission spectrum of fog in the infrared region of the spectrum (Contribution to the Tenth All-Union Conference on Spectroscopy, July 1956), Fiz. Sb. L'vovsk. Gos. Univ. (3):121-125 (1957), bibliography: 9 titles.

Markov, M. N., Comparison of low-inertia receivers of infrared radiation, Opt. i Spektroskopiya 3(2):158-161 (1957), bibliography: 16 titles.

Markov, M. N., Spectral sensitivity of low-inertia sputtered bolometer (Contribution to the Tenth All-Union Conference on Spectroscopy, July 1956), Fiz. Sb. L'vovsk. Gos. Univ. (3):403-405 (1957).

Motulevich, G. P., and Shubin, A. A., Determination of the optical constants of metals in the infrared, Opt. i Spektroskopiya 2(5):633-636 (1957), bibliography: 6 titles.

Motulevich, G. P., and Fabelinskii, I. L., Optical method of the absolute calibration of acoustic radiators at low sonic frequencies (Letter to the Editor), Akust. Zh. 3(2):205-206 (1957), bibliography: 3 titles.

Petrash, G. G., and Rautian, S. G., Precision of measuring optical density (Contribution to the Tenth All-Union Conference on Spectroscopy, July 1956), Fiz. Sb. L'vovsk. Gos. Univ (3):102-106 (1957), bibliography: 8 titles.

Podlovchenko, R. I., and Sushchinskii, M. M., Application of electronic computers to the calculation of molecular vibration frequencies (Contribution to the Tenth All-Union Conference on Spectroscopy, July 1956), Fiz. Sb. L'vovsk. Gos. Univ. (3):99-102 (1957), bibliography: 8 titles.

Podlovchenko, R. I., and Sushchinskii, M. M., Calculation and interpretation of vibration spectra of rotatory isomers of normal butane, Opt. i Spektroskopiya 2(1):49-61 (1957), bibliography: 13 titles.

Rakov, A. V., Variation of the Raman spectrum line breadth with the state of aggregation of the substance (Contribution to the Tenth All-Union Conference on Spectroscopy, July 1956), Fiz. Sb. L'vovsk. Gos. Univ. (3):229-230 (1957).

Rautian, S. G., On the theory of the echellette, Opt. i Spektroskopiya 2(2):279-280 (1957).

Rautian, S. G., and Petrash, G. G., Precision of measuring narrow absorption lines with elimination of the apparatus function (Contribution to the Tenth All-Union Conference on Spectroscopy, July 1956), Fiz. Sb. L'vovsk. Gos. Univ. (3):107-111 (1957).

Serdobol'skii, V. I., Application of dispersion relations for the determination of bound states, Zh. Eksperim. i Teor. Fiz. 33(5):1268-1270 (1957), bibliography: 1 title.

Sokolovskaya, A. I., and Bazhulin, P. A., Effect of temperature on the Raman spectrum in liquids (Contribution to the Tenth All-Union Conference on Spectroscopy, July 1956), Fiz. Sb. L'vovsk. Gos. Univ. (3):225-227 (1957), bibliography: 11 titles.

Sushchinskii, M. M., and Bazhulin, P. A., Application of the Raman effect to the study of composition and structure in materials, Usp. Fiz. Nauk 63(2):301-321 (1957), bibliography: 56 titles.

Velichkina, T. S., Mikheeva, L. F., and Yakovlev, I. A., Molecular scattering of light on phase transformations in a solid (Contribution to the Tenth All-Union Conference on Spectroscopy, July 1956), Fiz. Sb. L'vovsk. Gos. Univ. (3):111-115 (1957), bibliography: 10 titles.

Yakovlev, I. A., and Velichkina, T. S., Two new phenomena during phase transformations of the second kind, Usp. Fiz. Nauk 63(2):411-433 (1957), bibliography: 25 titles.

1958

Bogdanov, V. D., and Sushchinskii, M. M., Coefficients of anharmonicity and resonance interaction of the internal vibrations of CH group (Contribution to the Eleventh All-Union Conference on the Theory of Spectroscopy, Moscow, December 1957), Izv. Akad. Nauk SSSR, Ser. Fiz. 22(9):1067 (1958), bibliography: 4 titles.

Fabelinskii, I. L., Molecular scattering of light in liquids (Dissertation in pursuance of the Degree of Doctor of Physicomathematical Science), Tr. Fiz. Inst. Akad. Nauk 9:182-312 (1958), bibliography: 170 titles.

Fabelinskii, I. L., On the nature of visible sound waves, Priroda (3):127 (1958).

Malyshev, V. I., and Murzin, V. N., Study of the hydrogen bond in glycols and catechols (Contribution to the Eleventh All-Union Conference on the Theory of Spectroscopy, December, 1957), Izv. Akad. Nauk SSSR, Ser. Fiz. 22(9):1107-1108 (1958).

Motulevich, G. P., and Shubin, A. A., Role of interelectron collisions in metals in the infrared spectrum (Letter to the Editor), Zh. Eksperim. i Teor. Fiz. 34(3):757-758 (1958), bibliography: 5 titles.

Pesin, M. S., and Fabelinskii, I. L., Dispersion of the velocity of sound and propagation of hypersound in liquids, Dokl. Akad. Nauk SSSR 122(4):575-577 (1958), bibliography: 14 titles.

Petrash, G. G., Effect of measuring errors on the resolving power of infrared spectrometers, Inzh.-Fiz. Zh. 1(9):74-81 (1958), bibliography: 6 titles.

Petrash,G. G., and Rautian, S. G., Optimum conditions for measuring optical density on reduction to the ideal apparatus, Inzh.-Fiz. Zh. 1(11):80-91 (1958), bibliography: 10 titles.

Petrash, G. G., and Rautian, S. G., Allowance for apparatus distortions and characteristics of infrared spectrometers, Inzh.-Fiz. Zh. 1(7):61-71 (1958), bibliography: 22 titles.

Rautian, S. G., Practicable spectral apparatuses, Usp. Fiz. Nauk 66(3):475-517 (1958), bibliography: 130 titles.

Shelepin, L. A., Theory of particles with high spins, Zh. Eksperim. i Teor. Fiz. 34(6):1574-1586 (1958), bibliography: 13 titles.

Sobel'man, I. I., Some questions in the theory of the breadth of spectral lines (Dissertation in pursuance of the Degree of Master of Physicomathematical Science), Fiz. Inst. Akad. Nauk 9:313-359 (1958), bibliography: 52 titles.

Sobel'man, I. I., Quantum-mechanical theory of the intensity of Raman lines (Contribution to the Eleventh All-Union Conference on the Theory of Spectroscopy, December 1957), Izv. Akad. Nauk SSSR, Ser. Fiz. 22(9):1026-1029 (1958), bibliography: 8 titles.

Sobel'man, I. I., Breadth of spectral lines as a result of collisions with electrons, Fiz. Sb. L'vovsk. Gos. Univ. (4):303-305 (1958), bibliography: 10 titles.

Sobel'man, I. I., and Feinberg, E. L., Some optical effects of plasma oscillations in a solid, Zh. Eksperim. i Teor. Fiz. 34(2):494-500 (1958), bibliography: 8 titles.

Sobel'man, I. I., and Feinberg, E. L., Optical effects of collective electrons in metals (Contribution to the Eleventh All-Union Conference on the Theory of Spectroscopy, December 1957), Izv. Akad. Nauk SSSR, Ser. Fiz. 22(6):654-658 (1958), bibliography: 13 titles.

Sokolovskaya, A. I., and Bazhulin, P. A., Study of the temperature dependence of Raman line intensities (Contribution to the Eleventh All-Union Conference on the Theory of Spectroscopy, December 1957), Izv. Akad. Nauk SSSR, Ser. Fiz. 22(9):1068-1072 (1958), bibliography: 12 titles.

Sushchinskii, M. M., Conference on molecular spectroscopy in London, February 1958; Vestn. Akad. Nauk SSSR (6):87-88 (1958).

Sushchinskii, M. M., Study of vibration spectra in the region of the valence oscillations of CH (Contribution to the Eleventh All-Union Conference on the Theory of Spectroscopy, December 1957), Izv. Akad. Nauk SSSR, Ser. Fiz. 22(9):1063-1067 (1958), bibliography: 4 titles.

Sushchinskii, M. M., Application of rotation spectra of the Raman effect to determine structural parameters of molecules, Usp. Fiz. Nauk 65(3):441-450 (1958), bibliography: 26 titles.

Velichkina, T. S., Molecular light scattering in viscous liquids and amorphous solids (Dissertation in pursuance of the Degree of Master of Physicomathematical Science), Tr. Fiz. Inst. Akad. Nauk 9:59-124 (1958), bibliography: 50 titles.

Zaitsev, V. P., Motulevich, G. P., and Fabelinskii, I. L., Construction and absolute calibration of a magneto-electric acoustic radiator, Akust. Zh. 4(2):137-142 (1958), bibliography: 4 titles.

1959

Ablekov, V. K., On the analysis of spectrograms obtained on the Fabry—Perot Interferometer, Opt. i Spektroskopiya 6(4):562-564 (1959), bibliography: 6 titles.

Ablekov, V. K., and Fabelinskii, I. L., Spectral investigation of light scattered by viscous liquids and amorphous solids, Dokl. Akad. Nauk SSSR 125(2):297-299 (1959), bibliography: 12 titles.

Bazhulin, P. A., Malyshev, V. I., and Sushchinskii, M. M., Papers of G. S. Landsberg in the field of molecular spectroscopy, Collection: Studies in Experimental and Theoretical Physics, In Memory of Grigorii Samuilovich Landsberg, Moscow, Izd. An SSSR (1959), pp. 17-26.

Bazhulin, P. A., and Osipova, L. P., Differences of energy of rotation isomers in liquid and gaseous 1,2-fluorochlorethane, Opt. i Spektroskopiya 6(5):625-630 (1959), bibliography: 13 titles.

Bazhulin, P. A., and Smirnov, V. N., Study of the temperature dependence of the intensity of infrared absorption bands in liquids, Opt. i Sepktroskopiya 6(6):745-753 (1959), bibliography: 26 titles.

Bazhulin, P. A., and Sokolovskaya, A. I., Study of Raman line breadth as a function of temperature, Collection: Studies on Experimental and Theoretical Physics. In Memory of Grigorii Samuilovich Landsberg, Moscow, Izd. AN SSSR (1959), pp. 56-61, bibliography: 15 titles.

Bazhulin, P. A., and Sushchinskii, M. M., Methods of measuring Raman line intensities, Usp. Fiz. Nauk 68(1):135-146 (1959), bibliography: 17 titles.

Fabelinskii, I. L., Fine structure of the Rayleigh light scattering line in gases, Collection: Studies in Experimental and Theoretical Physics. In Memory of Grigorii Samuilovich Landsberg, Moscow, Izd. AN SSSR (1959), pp. 254-260, bibliography: 11 titles.

Landsberg, G. S., My first impressions of the demonstrations of L. F. Usagin, in: Ivan Filippovich Usagin (1855-1909), Moscow, Izd. Mosk. Gos. Univ., pp. 139-141.

Malyshev, V. I., and Murzin, V. N., Study of the hydrogen bond in substances the molecules of which contain two hydroxyl groups, Collection: Studies in Experimental and Theoretical Physics. In Memory of Grigorii Samuilovich Landsberg, Moscow, Izd. AN SSSR (1959), pp. 134-148, bibliography: 18 titles.

Malyshev, V. I., and Rautian, S. G., Vacuum two-beam diffraction spectrometer for the infrared (Contribution to the Twelvth All-Union Conference on Spectroscopy, November 1958), Izv. Akad. Nauk SSSR, Ser. Fiz. 23(10):1237-1239 (1959), bibliography: 5 titles.

Malyshev, V. I., and Rautian, S. G., Use of the echellette for large angles of diffraction, Opt. i Spektroskopiya 6(4):550-555 (1959), bibliography: 14 titles.

Markov, M. N., and Lindstrem, I. S., Optical properties of sputtered bismuth in the spectral range 3 to 15 μ, Opt. i Spektroskopiya 7(3):349-354 (1959), bibliography: 14 titles.

Markova, S. V., and Bazhulin, P. A., Determination of the infrared absorption coefficient of the CH_2 groups in cyclic compounds (Contribution to the Twelvth All-Union Conference on Spectroscopy, November 1958), Izv. Akad. Nauk SSSR, Ser. Fiz. 23(10):1186-1188 (1959), bibliography: 8 titles.

Mikhailov, G. V., Effect of pressure and temperature on the Raman spectrum of nitrogen, Zh. Eksperim. i Teor. Fiz. 36(5):1368-1373 (1959), bibliography: 8 titles.

Motulevich, G. P., Connection between the optical constants of metals and their microcharacteristics, Zh. Eksperim. i Teor. Fiz. 37(6):1770-1774 (1959), bibliography: 8 titles.

Motulevich, G. P., Fabelinskii, I. L., and Yakovlev, I. A., Works of G. S. Landsberg on the classical scattering of light. Collection: Studies in Experimental and Theoretical Physics. In Memory of Grigorii Samuilovich Landsberg, Izd. AN SSSR (1959), pp. 3-16.

Pesin, M. S., and Fabelinskii, I. L., Fine structure of Rayleigh line and propagation of hypersound in liquids of large viscosity, Dokl. Akad. Nauk SSSR 129(2):299-302 (1959).

Petrash, G. G., Choice of scanning rate, optimum time constant, and slit widths for spectrometer measurements, Opt. i Spektroskopiya 6(6):792-797 (1959), bibliography: 10 titles.

Podlovchenko, R. I., Sverdlov, L. M., and Sushchinskii, M. M., Vibration spectra and rotation isomerism of 2,3-dimethylbutane, Opt. i Spektroskopiya 6(2):146-153 (1959), bibliography: 9 titles.

Rakov, A. V., Effect of intermolecular action on Raman effect line breadth in liquids, Opt. i Spektroskopiya 7(2):202-207 (1959), bibliography: 16 titles.

Rautian, S. G., On the theory of the echellette, Opt. i Spektroskopiya 7(4):564-566 (1959), bibliography: 9 titles.

Savin, F. A., and Sobel'man, I. I., Intensities in Raman spectra and metallic model of the molecule, I, Opt. i Spektroskopiya 7(6):733-739 (1959), bibliography: 15 titles.

Savin, F. A., and Sobel'man, I. I., Intensities in Raman spectra and metallic model of the molecule, II, Opt. i Spektroskopiya 7(6):740-743 (1959), bibliography: 7 titles.

Serdobol'skii, V. L., Theory of scattering in terms of quasistationary states, Zh. Eksperim. i Teor. Fiz. 36(6):1903-1908(1959), bibliography: 6 titles.

Shelepin, L. A., Theory of relativistically invariant equations, Zh. Eksperim. i Teor. Fiz. 37(6):1626-1638 (1959), bibliography: 7 titles.

Smirnov, V. N., Study of the temperature dependence of the intensity of infrared absorption bands in solutions, Opt. i Spektroskopiya 7(4):472-477 (1959), bibliography: 21 titles.

Smirnov, V. N., and Bazhulin, P. A., Study of the temperature dependence of the intensity of infrared absorption bands in gases, Opt. i Spektroskopiya 7(2):193-201 (1959), bibliography: 36 titles.

Sobel'man, L. L., Quantum-mechanical theory of Raman effect line intensities, Collection: Studies in Experimental and Theoretical Physics. In Memory of Grigorii Samuilovich Landsberg, Moscow, Izd. AN SSSR (1959), pp. 192-210, bibliography: 19 titles.

Sokolovskaya, A. L., and Rautian, S. G., Effect of the refractive index of the scattering medium on the Raman effect line intensity, Opt. i Spektroskopiya 6(1):52-54 (1959), bibliography: 10 titles.

Studies in Experimental and Theoretical Physics. In Memory of Grigorii Samuilovich Landsberg, Edited by Dr. of Physicomathematical Science L. L. Fabelinskii, Moscow, Izd. AN SSSR (1959), 304 pp. (Academy of Sciences USSR, P. N. Lebedev Physical Institute.)

Sushchinskii, M. M., Raman effect line breadths from the anisotropy of the polarizability derivative tensor, Collection: Studies in Experimental and Theoretical Physics. In Memory of Grigorii Samuilovich Landsberg, Moscow, Izd. AN SSSR (1959), pp. 211-217, biliography: 8 titles.

Zubov, V. A., Petrash, G. G., and Sushchinskii, M. M., Two-beam spectrometer for studying Raman spectra, Pribory i Tekhn. Eksperim.(5):119-120 (1959), bibliography: 2 titles.

Zubov, V. A., Petrash, G. G., and Sushchinskii, M. M., Some applications of a high-dispersion spectrometer for molecular analysis from Raman spectra (Contribution to the Twelfth All-Union Conference on Spectroscopy, November, 1958), Opt. i Spektroskopiya 6(6):827-829 (1959), bibliography: 5 titles.

1960

Ablekov, V. K., Pesin, M. S., and Fabelinskii, L. L., Realization of a medium with negative absorption coefficient (Letter to the Editor), Zh. Eksperim. i Teor. Fiz. 39(3):892-893 (1960), bibliography: 8 titles.

Bazhulin, P. A., and Lazarev, Yu. A., Study of Raman spectra in gases at low pressure by a photoelectric method, Opt. i Spektroskopiya 8(2):206-213 (1960), bibliography: 34 titles.

Golovashkin, A. L., Motulevich, G. P., and Shubin, A. A., Measurements of the optical constants of metals at low temperatures, Pribory i Tekhn. Eksperim. (5):74-76 (1960), bibliography: 4 titles.

Golovashkin, A. L., Motulevich, G. P., and Shubin, A. A., Determination of the microcharacteristics of aluminum from measurements of optical constants and specific conductivity, Zh. Eksperim. i Teor. Fiz. 38(1):51-55 (1960), bibliography: 14 titles.

Markov, M. N., and Khaikin, A. S., Optical properties of massive bismuth in the spectral range 3-36 μ, Opt. i Spektroskopiya 9(4):487-492 (1960), bibliography: 18 titles.

Markov, M. N., and Krugliakov, E. P., Zonal sensitivity of PbS photoresistances, Opt. i Spektroskopiya 9(4):538-540 (1960), bibliography: 5 titles.

Naberukhin, Yu. L., and Sushchinskii, M. M., Second-order lines in hydrocarbon spectra, Opt. i Spektroskopiya 9(5):576-581 (1960), bibliography: 17 titles.

Pesin, M. C., and Fabelinskii, L. L., Spectroscopical investigation of the propagation of hypersound oscillations in viscous liquids, Dokl. Akad. Nauk SSSR 135(5):1114-116 (1960), bibliography: 6 titles.

Petrash, G. G., Effect of scanning and choice of optimum measuring conditions, Opt. i Spektroskopiya 8(1):122-123 (1960), bibliography: 2 titles.

Petrash, G. G., Optimum conditions of measuring absorption line widths, Opt. i Spektroskopiya 9(3):423-424 (1960), bibliography: 3 titles.

Petrash, G. G., Breadth and shape of infrared absorption bands, Opt. i Spektroskopiya 9(1):121-123 (1960), bibliography: 7 titles.

Rautian, S. G., and Sobel'man, L. L., On negative absorption in metal vapors (Letter to the Editor), Zh. Eksperim. i Teor. Fiz. 39(1):217-219 (1960), bibliography: 9 titles.

Serdobol'skii, V.L., Dispersion formulas for overlapping levels, Zh. Eksperim. i Teor. Fiz. 38(6):1903-1906 (1960), bibliography: 4 titles.

Serdobol'skii, V.I., Dispersion formulas in the theory of nuclear reactions, Nucl. Phys. 21(2):245-255 (1960).

Sokolovskaya, A. I., Temperature dependence of the intensity of Raman lines in vapor and liquid phases, Opt. i Spektroskopiya (5):582-586 (1960), bibliography: 23 titles.

Sokolovskaya, A. I., and Bazhulin, P. A., Temperature dependence of the intensity of Stokes and anti-Stokes lines in the Raman effect of liquids, Opt. i Spektroskopiya 8(3):394-397 (1960), bibliography: 17 titles.

Suchchinskii, M. M., Raman spectra and the structure of hydrocarbons (Dissertation in pursuance of the Degree of Doctor of Physicomathematical Science), Tr. Fiz. Inst. Akad. Nauk 12:54-224 (1960), bibliography: 180 titles.

Vainshtein, L. A., and Sobel'man, I. I., Derivation of radial equations in the theory of electron-atom collisions, Zh. Eksperim. i Teor. Fiz. 39(3):767-775 (1960), bibliography: 5 titles.

Vainshtein, L. A., and Sobel'man, I. I., Methods of calculating the effective cross sections for the excitation of atoms by electrons (Contribution to the first All-Union Conference on Electronic and Ionic Collisions, 1959), Izv. Akad. Nauk SSSR, Ser. Fiz. 24(8):943-945 (1960), bibliography: 7 titles.

1961

Ablekov, V. K., Zaitsev, V. P., and Pesin, M. S., Mercury-zinc and mercury-cadmium high-intensity lamps, Pribory i Tekhn. Eksperim. (2):140-142 (1961), bibliography: 13 titles.

Kuznetsova, T. I., and Rautian, S. G., Theory of quantum generators, Zh. Eksperim. i Teor. Fiz. 43(5):1897-1903 (1961), bibliography: 4 titles.

Petrash, G. G., Breadth and shape of infrared absorption bands, Opt. i Spektroskopiya 9(1):121-123 (1961), bibliography: 7 titles.

Rakov, A. V., Temperature dependence of the Raman effect line breadths for substances in the crystalline state, Opt. i Spektroskopiya 10(6):713-716 (1961), bibliography: 6 titles.

Rautian, S. G., and Sobel'man, I. I., On the question of negative absorption, Opt. i Spektroskopiya 10(1):134-135 (1961), bibliography: 6 titles.

Rautian, S. G., and Sobel'man, I. I., Line shape and dispersion in the region of the absorption band allowing for forced transitions, Zh. Eksperim. i Teor. Fiz. 41(2):456-464 (1961), bibliography: 9 titles.

Rautian, S. G., and Sobel'man, I. I., Photodissociation of molecules as a means of obtaining media with negative absorption coefficients, Zh. Eksperim. i Teor. Fiz. 41(6):2018-2020 (1961), bibliography: 8 titles.

Serdobol'skii, V. I., Dispersion formulas allowing for optical interaction, Zh. Eksperim. i Teor. Fiz. 40(2):590-596 (1961), bibliography: 5 titles.

Shelepin, L. A., Racah method in the theory of relativistic equations, Zh. Eksperim. i Teor. Fiz. 40(5):1369-1383 (1961), bibliography: 8 titles.

Sokolovskaya, A. I., Intensity of Raman lines in the liquid and vapor phases of matter, Opt. i Spektroskopiya 11(4):478-481 (1961), bibliography: 17 titles.

1962

Ablekov, V. K., Some experimental studies of the induced radiation of a mixture of gases, Zh. Eksperim. i Teor. Fiz. 42(3):736-739 (1962), bibliography: 5 titles.

Aref'ev, I. M., and Malyshev, V. I., Study of the hydrogen bond of hydrogen halides, Opt. i Spektroskopiya 13(2):206-211 (1962), bibliography: 15 titles.

Bazhulin, P. A., Study of rotational and rotational-vibrational gas spectra by the Raman effect method, Usp. Fiz. Nauk 77(4):639-648 (1962), bibliography: 14 titles.

Bazhulin, P. A., Lazarev, Yu. A., and Desyatkova, N. V., Study of the intensity and degree of depolarization of the Raman spectrum vibrational lines of butadiene in the gaseous phase, Opt. i Spektroskopiya 13(1):75-78 (1962), bibliography: 16 titles.

Fabelinskii, I. L., Some results of an experimental study of the fine structure of Rayleigh light-scattering lines in liquids of various viscosity, Usp. Fiz. Nauk 77(4):649-662 (1962), bibliography: 43 titles.

Golovashkin, A. I., and Motulevich, G. P., Sensitive low-inertia thermoresistance for helium temperatures, Pribory i Tekhn. Eksperim. (2):182-185 (1962), bibliography: 5 titles.

Lazarev, Yu. A., Broadening of lines in the rotational and rotational-vibrational Raman spectra in the gas phase, Opt. i Spektroskopiya 13(5):655-662 (1962), bibliography: 18 titles.

Naberukhin, Yu. I., Effect of the geometry of the illumination on the observed intensity and degree of depolarization of Raman lines, Opt. i Spektroskopiya 13(4):498-504 (1962), bibliography: 17 titles.

Rakov, A. V., Temperature dependence of infrared absorption bandwidths, Opt. i Spektroskopiya 13(3):369-373 (1962), bibliography: 10 titles.

Serdobol'skii, V. I., Forced radiation in a strong electromagnetic field, Zh. Eksperim. i Teor. Fiz. 43(6):2121 (1962), bibliography: 7 titles.

Shelepin, L. A., Covariant theory of relativistic wave equations, Nucl. Phys. 33(4):580-593 (1962), bibliography: 13 titles.

Sushchinskii, M. M., Molecular spectroscopy, Vestn. Akad. Nauk SSSR (10):36-45 (1962).

Sushchinskii, M. M., and Zubov, V. A., On the connection between Raman spectra and electronic absorption, Opt. i Spektroskopiya 13(6):766-774 (1962), bibliography: 10 titles.

Vil'ner, L. D., Rautian, S. G., and Khaikin, A. S., On certain possibilities of using the Fabry—Perot interferometer with internal illumination, Opt. i Spektroskopiya 12(3):437-439 (1962), bibliography: 3 titles.

Zubov, V. A., Study of the relation between the degree of depolarization of Raman lines and the frequency of the exciting light, Opt. i Spektroskopiya 13(6):861-862 (1962), bibliography: 6 titles.